From King Cane to the Last Sugar Mill

T0243264

FROM KING CANE TO THE LAST SUGAR MILL

Agricultural Technology and the Making of Hawai'i's Premier Crop

C. ALLAN JONES & ROBERT V. OSGOOD

University of Hawai'i Press

HONOLULU

Library of Congress Cataloging-in-Publication Data

Jones, C. Allan, author.
  From king cane to the last sugar mill : agricultural technology and the
making of Hawai'i's premier crop / C. Allan Jones and Robert V. Osgood.
    pages cm
  Includes bibliographical references and index.
  ISBN 978-0-8248-4000-6
  1. Sugarcane industry—Hawaii—History.   2. Sugarcane—Technological
innovations—Hawaii.   I. Osgood, Robert V., author.   II. Title.
  HD9107.H3J65 2015
  338.1'736109969—dc23

                                                                2014033715

  ISBN 978-0-8248-9576-1 (paperback)

CONTENTS

FOREWORD

Hawai'i is a magical place. Worldwide, the name Hawai'i conjures images of diverse natural scenery, coconut tree–edged sandy beaches, active volcanoes, and an exotic mix of people, cultures, and customs spanning Polynesia, Asia, and Europe. Hawai'i is geologically young and one of the last places on Earth to be populated by humans. According to archaeological evidence, it was first settled by Polynesians between 300–500 CE. The exact dates are not known since there is no recorded history of Hawai'i before the first European contact by Captain James Cook in 1778. Pre-European history of Hawai'i was conveyed solely by oral chants until the native language could be transcribed. The lack of a written early history adds to Hawai'i's mystery and magic, which in turn have helped make it attractive to many writers eager to elucidate various aspects of its fascinating story.

One of the more interesting features of Hawai'i's history, explored in this book, concerns the dominant role played by the sugarcane industry in shaping (some critics might say "controlling") nearly every feature of the state's current character, especially its unique mix of Polynesian culture and the cultures and customs brought to the islands by immigrants imported to labor on sugarcane plantations. The industry was once so dominant, in fact, that it has been referred to in more than one book title, including this one, as "King Cane."

Sugarcane was a defining force in Hawai'i from the time of the first humans' arrival. Archaeological evidence and oral histories suggest that sugarcane was introduced to Hawai'i by Polynesians more than a thousand years before Captain Cook recorded it in his logs. The economics of producing sugarcane in Hawai'i helped drive the Westernization of Hawaii for much of the nearly two and a half centuries since Cook's landing. Unfortunately, from the perspective of this book's authors, and mine as a sugarcane enthusiast, our generation may be seeing sugarcane's demise in Hawai'i. Only one sugarcane production company remains, Hawaiian Commercial & Sugar

Co. Ltd. on the island of Maui, the "last sugar mill" mentioned in the book's title, and its survival is not guaranteed.

Some might be tempted to dismiss this book as a public relations release intended to influence company or government policy about water and agriculture or as a technical bulletin about sugarcane agronomic practices. To do so would be a mistake. Although it is true that this text focuses on the history of water management and a single sugarcane production company in an island state, it takes a broad approach, putting these issues in the context of Hawai'i and world politics, economics, labor management, and the technological advances in agriculture that became available in particular eras to improve crop production.

The authors are agricultural scientists who once worked for the Experiment Station of the Hawaiian Sugar Planters' Association (HSPA), giving them extensive firsthand experience with the technological and scientific advances that allowed Hawai'i's sugar industry to lead the world's sugar production throughout most of the twentieth century. They also had access to HSPA's extensive archival materials (photographs, memos, letters, etc.) and to personal contacts with families and workers in the sugarcane industry. The archival materials and the contacts with former industry workers, combined with the authors' own fascination with sugarcane, add rare personal touches to this historical account.

The present book is a scholarly, highly informative, and entertaining history of the rise and fall of Hawai'i's sugarcane industry from before the first sugarcane plantation in 1835 up to the present. The authors describe how Hawai'i's sugar industry has been forced to change and evolve repeatedly in response to international crises and U.S. policies on sugar prices, commerce, labor, and environmental concerns. The book is highly recommended to students of Hawaiiana and those seeking to understand how twenty-first–century Hawai'i came to be.

Paul H. Moore, Ph.D.
Hawaii Agriculture Research Center, Kunia, HI

PREFACE

King Cane and the Last Sugar Mill

In 1939, when John Vandercook wrote the original *King Cane, the Story of Sugar in Hawaii,* there were thirty-eight sugar plantations operating in Hawai'i on four islands. Now there is only one—Hawaiian Commercial & Sugar Company (HC&S), a subsidiary of Alexander & Baldwin, located on Maui. It is the origin of our title: *From King Cane to the Last Sugar Mill.* This is not a comprehensive review of the Hawaiian sugar industry but rather a look at how events and technology shaped the industry, with an emphasis on the development and evolution of HC&S against a background of politics, commerce, science, and technology. Our book traces the history of Hawai'i's sugarcane from its Polynesian origins in the Kingdom of Hawai'i, through the annexation to the United States as a territory, under statehood, and finally to the last commercial sugar mill on Maui, which is owned by HC&S. We emphasize the many factors that have affected the Hawaiian sugar industry and HC&S, including technology, politics, economics, social issues, and cooperation among the plantations and the agencies servicing the industry. If it had not been responsive to these factors, the sugar industry would have been short-lived, and the last mill would have closed long ago.

Five chapters designated by significant events and dates are used to trace sugar industry history:

    I. Birth of an Industry—to 1875
    II. Sugar Booms—1876 to 1897
  III. Industry Growth and Labor Unrest—1898 to 1929
   IV. Depression, War, Federal Legislation, Science, and Technology—1930 to 1969
    V. Drip Irrigation and New Disease Resistant Varieties Save HC&S—1970 to 2014

Our book emphasizes that the sugar industry in Hawai'i developed differently than the industry did in other places around the world, where slaves provided labor and where there was little concern for the health and well-being of the workforce. Also in contrast to other sugarcane industries, the crop in Hawai'i was grown for up to two or more years before harvest, and a very long milling season was possible owing to mild weather. This required a year-round supply of reliable and well-trained labor for all aspects of sugarcane culture and milling. Labor was sourced primarily from Asia and Europe and was brought to Hawai'i under labor contracts and in compliance with U.S. law. Since labor was almost always scarce on the islands, the sugarcane workforce in Hawai'i was well paid. However, economic pressures forced the industry to reduce labor costs with labor-saving technologies, ultimately resulting in one of the most mechanized sugar industries in the world.

Because the culture of sugarcane in Hawai'i was unique, the industry was a leader in developing advanced technologies, including specialized agricultural implements, water harvesting and transport, in-field irrigation systems, the fine tuning of fertilizer application, mechanical harvesting, cane loading and transport, and factory operations. The two-year crop cycle also required that unique varieties of sugarcane be bred.

Throughout its history, the Hawai'i sugar industry took full advantage of the islands' abundant water resources, first by collecting runoff from the wet windward slopes and transferring it in ditches to the sunny but dry leeward sides of the islands. Later, the industry dug wells to pump from the lens of freshwater resting atop seawater under plantation cane fields. For decades, the industry, often led by HC&S, improved its irrigation methods and technologies, eventually developing some of the most extensive and advanced drip irrigation systems in the world. This highly efficient technology allowed the islands' irrigated plantations to continue producing well beyond the time when they would have been forced to close had they not made the change. Indeed, HC&S remains the world's largest drip-irrigated sugar plantation.

Our book jacket is a drawing depicting early 1940 sugar plantation operations from soil preparation to the factory; the image was originally published in a Hawaiian Sugar Planters' Association (HSPA) public relations pamphlet titled "The Story of Sugar in Pictures" and is used here by permission of the Hawaii Agriculture Research Center (HARC), the successor of the HSPA.

The authors acknowledge Stephanie Whalen, executive director of HARC, who granted permission to use photographs and drawings sourced from HARC files; Ann Marsteller, HARC librarian, who located resource materials in the files; Roslyn Lightfoot of the Alexander & Baldwin Sugar Museum for use of historical materials; Paul and Melinda Moore, who reviewed the manuscript and prepared the foreword to the book; Jaehak Jeong, who provided a map of the Maui irrigation system; and the U.S. Department of Agriculture and the Office of Naval Research, for their financial assistance under account agreement 108-6206-110. We also thank Mae Nakahata at HC&S for early help with the project.

INTRODUCTION

Sugarcane has been grown in Hawai'i since the arrival of its Polynesian settlers, and commercial sugar production has been an economic force on the islands since the mid-1800s. Led by the Hawaiian Sugar Planters' Association (HSPA), the Hawaiian Islands grew to be a world leader in sugarcane production and sugar technology in the early and mid-twentieth century. The industry's many sugar companies voluntarily supported the Experiment Station of the HSPA, which was staffed with the world's sugarcane research leaders throughout most of the twentieth century. However, after more than one hundred years of almost continuous growth and development, the industry faltered in the last quarter of the twentieth century. Sugar growers were unable to cope with continually increasing costs of production that outpaced domestic and international sugar prices. Of the twenty-eight sugar companies operating on the islands at the beginning of 1970, sixteen remained in 1980, twelve were still in production at the beginning of 1990, and only four survived until January 2000. By 2011, just one—Hawaiian Commercial & Sugar Company, better known for over a century as HC&S—was still in operation.

It is no accident that HC&S was the only Hawaiian sugar company to survive the economic challenges that began in the 1970s. A century earlier, using irrigation water gathered from the rainy windward slopes of East Maui, the company began its ongoing work to improve cane fields in warm and sunny Central Maui. It merged with or absorbed the best lands of smaller companies, mechanized its field operations, enhanced its irrigation technologies, and merged and modernized its factories, all while decreasing the size and increasing the efficiency of its labor force. The result was a consistently profitable company, farming mostly its own land throughout the late nineteenth and most of the twentieth centuries. In recent years HC&S has been by far the largest component of the storied Alexander & Baldwin Agribusiness operation, and with the exception of operating losses in 2008 and 2009,

A&B's integrated operations were consistently profitable through 2013. However, declining prices since 2012 threaten future profitability of traditional agribusiness operations focused on the production and sale of raw sugar, molasses, and bagasse-generated electricity. As HC&S faces an uncertain future, we look back on the evolution of the Hawai'i sugar industry, with particular emphasis on the industry on Maui and the development of HC&S.

As agricultural scientists, our emphasis is on the technological and scientific advances that allowed Hawai'i's sugar industry to be among the world's leaders throughout most of the twentieth century. But we would be remiss if we did not acknowledge the enormous societal and environmental changes caused by the sugar industry's aggressive search for labor, land, and water resources. To address this broader perspective, we briefly describe the islands' system of land and water rights, indigenous Hawaiian agricultural practices, and early attempts to diversify the Kingdom's agriculture. We also discuss the effects of external forces that stimulated Hawaiian agricultural development, including demand created by the California gold rush, the American Civil War, and the Reciprocity Treaty with the United States.

One of the most important factors in the development of the sugar industry was the declining population and political influence of the Hawaiian people. Until the 1870s Hawaiians and part-Hawaiians made up more than 80 percent of the sugar industry's small labor force. But European diseases reduced the Hawaiian population dramatically, from about 130,000 in 1832, to 71,000 in 1853, and just 34,000 in 1890. The industry grew rapidly after the Reciprocity Treaty of 1876 eliminated tariffs between the Kingdom and the United States, and the declining Hawaiian population could no longer provide the labor needed by the plantations. In response, beginning in the 1870s, the industry aggressively recruited immigrant labor, first from China and later from Japan, Portugal, Korea, and the Philippines. To manage their increasingly diverse labor force, the plantations developed strongly paternalistic labor management systems, providing housing and social services in plantation villages or "camps" that were often segregated by ethnic group. These forces also led to a rapid growth in the number of plantations, improvements in sugar factories, and the construction of extensive plantation irrigation systems that collected water from the rainy windward sides of the islands and carried it to the warm sunny leeward cane fields.

In the 1890s the leaders of the sugar industry played a major role in wresting political power away from the Hawaiian Kingdom. After Hawai'i became a territory in 1898, it fell under U.S. law, which excluded Chinese immigration and invalidated the Kingdom's contract labor laws, greatly reducing the plantations' hold on labor. Japanese workers began to assert their rights, calling a major strike against Oʻahu plantations in 1909. In response, the industry promoted immigration from the Philippines, and by the 1920s Japanese and Filipino workers made up almost 80 percent of the industry's labor force. Sensing their strength, Japanese and Filipino workers cooperated to organize a major strike in 1919–1920, which led to wage and other concessions by the industry.

The early decades of the twentieth century also saw major improvements in industry irrigation systems and the development of wells to tap the islands' rich groundwater resources. The Experiment Station of the HSPA became a powerful force for scientific agricultural practices, promoting the use of improved sugarcane varieties, better irrigation systems, the application of fertilizers, and biological control of pests. Plantations also expanded in size, using steam-powered locomotives to bring cane from their fields to their factories and steam engines to drive their mills and plows. Factory mills and other sugar-making equipment improved, and in the 1920s the industry was probably the most technologically advanced in the world.

Sugarcane survived the low sugar prices of the Great Depression and the labor shortages during World War II by continually mechanizing processes, which reduced labor costs and increased labor productivity. It enjoyed another period of science-driven gains in productivity and profitability in the 1950s and 1960s. But beginning in the 1970s, the industry faced numerous and more critical challenges. Major fluctuations in international sugar prices and the U.S. government price support program, caused in large part by a major increase in nonsucrose sweeteners, produced unprecedented economic pressures. Many of the state's less-efficient, mostly unirrigated plantations were forced out of business, reducing the number of plantations from twenty-seven in 1970 to fourteen in 1980. The more efficient irrigated plantations survived by converting from furrow to drip irrigation, which reduced costs and in some cases increased sugar yields. But economic pressures on the industry continued, and the number of plantations continued to fall to twelve in 1990 and only four in 2000. By 2011 HC&S was the lone plantation remaining in the state of Hawaiʻi.

Because of its large size, excellent surface and groundwater resources, efficient drip irrigation system, increasing mechanization and automation, and warm sunny climate, HC&S remained generally profitable into the 2000s. However, severe drought conditions and decreased pumping of groundwater reduced sugar yields in 2006–2009 and caused substantial operating losses in 2008 and 2009. While profits rebounded from 2010 through 2013 due to increased yields and higher sugar prices, great uncertainties remain. Groups concerned with environmental quality, Hawaiian water rights, and the assumed public health effects of pre-harvest burning have challenged HC&S in state courts. Thus far, the company's experiments with harvesting green cane to reduce smoke from pre-harvest burning (and produce more biomass to generate more electricity) have not produced successful alternatives. Nor has HC&S found economical methods of transforming sugarcane biomass, molasses, or sugar into transportation fuels such as ethanol or diesel. As of this writing, the future of HC&S, the last sugar company in Hawai'i, remained uncertain.

We strive in this book to offer a balanced view of the sugarcane industry in Hawai'i. Our goal is to help residents and visitors to Hawai'i and Maui appreciate the long history of the Hawaiian sugar industry and better understand the complex web of forces—scientific, technological, economic, environmental, and ethnic—it reflects.

1

BIRTH OF AN INDUSTRY—TO 1875

Hawaiian Land, Agriculture, and Irrigation

The cultivation of sugarcane in the Hawaiian Islands began with the arrival of the first humans, from the Marquesas Islands in the southern Pacific Ocean between 500 and 700 CE. There were no food plants available in Hawai'i when the Polynesian peoples arrived there, with the exception of some marine algae species and possibly coconut, making plant introductions critical for survival. Traveling for weeks in their double-hulled canoes, Polynesian voyagers brought with them pigs, chickens, and dogs, as well as about thirty useful "canoe plants" common throughout the South Pacific.[1] These included a few trees but were mainly food, medicinal, and fiber crops, most notably kalo (taro, *Colocasia esculenta*), 'uala (sweet potato, *Ipomoea batatas*), mai'a (banana, *Musa* spp.), wauke (paper mulberry, *Broussonetia papyrifera*), niu (coconut, *Cocos nucifera*), kō (sugarcane, *Saccharum officinarum*), 'ulu (breadfruit, *Artocarpus altilis*), pia (Polynesian arrowroot, *Tacca leontopetaloides),* 'ohe (bamboo, *Schizostachyum glaucifolium/Bambusa vulgaris)*, ipu (bottle gourd, *Lagenaria siceraria, L. vulgaris)*, noni (beach mulberry, *Morinda citrifolia*), 'ōhi'a 'ai (mountain apple, *Eugenia malaccensis)*, 'olena (turmeric, *Curcuma longa)*, olonā *(Touchardia latifolia)*, uhi (yam, *Dioscorea alata*), kī (ti, *Cordyline terminalis)*, kukui (candlenut, *Aleurites moluccana)*, and 'ape (elephant's ear, *Alocasia macrorrhiza)*. Several trees—among them hau (*Hibiscus tiliaceus*), kamani (*Calophyllum inophyllum*), and milo (*Thespesia populena*)—were introduced primarily for fiber and wood.

Kō (sugarcane, *Saccharum officinarum*) was important because it provided food, fiber, and medicine; it also served as a windbreak and a stabilizer for the berms surrounding flooded *lo'i* (kalo ponds). Up to fifty varieties of sugarcane may have been cultivated on the islands, with specific types used for different purposes. Which varieties were introduced by the Polynesians remains unclear; mutations of the original introductions apparently occurred, and later introductions were made by Westerners arriving on trading ships.

Sugarcane was prized by the Hawaiians for its flavor-enhancing and medicinal properties. Stalks were cut, peeled, and pounded. The sweet juice was then squeezed out and fed to babies, mixed with medicine to improve its taste, and used to sweeten the paste of pounded kalo, called poi, and haupia, a pudding made from arrowroot (pia) and coconut (niu). Young sugarcane shoots, along with salt and other herbs, were wrapped in ti (kī) leaves and baked, and the resulting juice was used for healing cuts. The fibrous portion of the cane stalk was used to clean teeth, and the leaves were used for thatching roofs.[2]

Although the Hawaiian Islands may have been visited by a Chinese junk in the thirteenth century CE and by Spanish ships 300 years later, the first documented visit by non-Polynesians was in 1778, when the islands were "discovered" by Captain James Cook of England. While in search of a northwestern sailing route from the Pacific to the Atlantic, Cook visited and traded with Hawaiians on the major islands of the Hawaiian chain. Kaua'i, the oldest of the islands, was Cook's first stop, and according to his detailed account of his travels, he filled his ship's water barrels at a spring, likely near present-day Waimea Town. Cook recorded his observations: "We saw no wood, but what was up in the interior part of the island, except a few trees about the villages; near which, also, we could observe several plantations of plantains and sugar-canes, and spots that seemed cultivated for roots."[3]

Describing the crops he saw, Cook noted that "the greatest part of the land as quite flat, with ditches full of water intersecting different parts, and roads that seemed artificially raised to some height. The interspaces were, in general, planted with *taro* [kalo], which grows here with great strength, as the fields are sunk below the common level, so as to contain the water necessary to nourish the roots." He further remarked, "On the drier spaces were several spots where the cloth-mulberry [wauke] was planted in regular rows; also growing vigorously, and kept very clean. The cocoa-trees [coconut] were not in so thriving a state, and were all low; but the plantain-tree [banana or mai'a] made a better appearance, though they were not large."[4]

After stopping on Kaua'i, Cook visited Maui, which he spelled phonetically as "Mowee." From his ship five miles off Maui's wet windward coast, he commented, "In the country was an elevated saddle hill [Haleakalā], whose summit appeared above the clouds. From this hill, the land fell in a gentle slope, and terminated in a steep, rocky coast, against which the sea broke in a dreadful surf." Unable to land because of the weather, Cook took his ship westward, observing "people on several parts of the shore, and some houses and plantations. The country seemed to be both well wooded and watered; and running streams were seen falling into the sea in various places." He also noted the isthmus between East and West Maui, and, welcoming visitors from Maui on board, traded nails and pieces of iron for cuttlefish. Cook reported that the natives brought few fruits and little taro with them "but told us that they had plenty of them on their island, [and] also hogs and fowls."[5]

Probably while on Kaua'i, Cook traded for enough sugarcane to brew sugarcane beer for his crew. The beer was ready by the time he arrived off Maui's windward coast. "But when the cask was now broached," Cook explained, "not one of my crew would even so much as taste it," because they feared for their health. Disgusted with the crew's reaction to the beer, Cook complained that "few commanders have introduced into their ships more novelties, as useful varieties of food and drink, than I have done. . . . [and] I have been able to preserve my people, generally speaking, from the dreadful distemper, the scurvy, which has perhaps destroyed more of our sailors, in their peaceful voyages, than have fallen by the enemy in military expeditions." He then gave orders that "no grog should be served" to the crew, but he and the officers "continued to make use of this sugar-cane beer, whenever [the] materials for brewing it" were available. According to Cook, the addition of "a few hops, of which we had some on board, improved it much. It has the taste of new malt beer; and I believe that no one will doubt its being very wholesome. And yet my inconsiderate crew alleged that it was injurious to their health."[6]

Cook's was the first of many visits by Europeans and Americans that soon began to change the face of the islands. Aside from "the venereal distemper" that his crew had transmitted to the islanders,[7] Cook introduced a pair of English pigs; oranges arrived in 1792, and Captain George Vancouver brought the first cattle from California in 1793 and 1794.[8] Fur traders soon stopped in the islands on their voyages between northwestern North America and China. Valuing their iron, firearms, and other goods from the West, the warring Hawaiian chiefs retained Westerners to build ships and help them wage war with Western weapons, quickly changing the nature of Hawaiian warfare. King Kamehameha I, a chief on the island of Hawai'i, conquered most of the islands in a series of battles between 1782 and 1810, culminating with Kaua'i's chief accepting Kamehameha as overlord and agreeing to pay him tribute. Upon Kamehameha's death in 1819, his first son, Liholiho

(Kamehameha II), took power. In short order, Liholiho, in alliance with Kamehameha I's favorite wife, Ka'ahumanu, overthrew many of the old *kapu*, or strict taboos, making way for more rapid societal change.[9]

Because the Hawaiian monarchy ruled from O'ahu in the early nineteenth century, most early visitors and immigrants lived there, but their general observations on the Hawaiian people, their agriculture, and their management of water resources could be extended to Maui and the other islands. Archibald Campbell, who lived on O'ahu in 1809 and 1810, provided an early account of Hawaiian society and agriculture from a Western perspective. He reported that only about 60 "whites" lived on O'ahu, a third of them Americans and the remainder mostly English, including six or eight convicts who had escaped from New South Wales. In an account of his travels first published in 1817, Campbell observed, "The chiefs are the proprietors of the soil, and let the land in small farms to the lower class, who pay them a rent in kind, generally pigs, cloth, or mats, at four terms a year."[10] Regard for the chiefs and their agents, as well as the elaborate kapu system, were used to maintain control of the people.[11]

Early non-Hawaiian visitors encountered few plants that were very appealing to Western tastes. As a result, the introduction of more familiar food crops was of great interest to the rapidly increasing non-Hawaiian population. Don Francisco de Paula Marin, an immigrant from Spain in the late eighteenth century, became an adviser to Kamehameha I and was credited with introducing or growing coffee, pineapple, grapes, avocado, mango, and other new crops. He took advantage of the frequent arrivals of whaling ships to receive plants for trial in Hawai'i and send island-grown plants to others around the world. Marin also produced brandy, rum, and wine, raising objections from Christian missionaries who opposed their consumption.

The native Hawaiian system of land nomenclature, management, and control had developed since Polynesians first occupied the islands. Each of the Hawaiian Islands was divided into several districts, known as *moku*. The traditional moku of Maui were Kā'anapali, Lahaina, Wailuku, Hāmākuapoko, Hāmākualoa, Ko'olau, Hāna, Kīpahulu, Kaupō, Kahikinui, Honua'ula, and Kula. Several of these names were later taken by sugar plantations, and many of them are still in use. Each moku was partitioned into several *ahupua'a*, each of which was usually a long narrow strip extending from the sea to the top of the principal mountain or ridge of the island. On the windward sides of the islands, the typical ahupua'a was a narrow valley bounded on each side by ridges. Naturally, some lands did not lend themselves to this type of natural division, and ahupua'a varied greatly in size and shape.[12]

Within an ahupuaʻa, tenants were granted home sites and kalo patches known as *kuleana*. All the chiefs and tenants held the land in trust and owed the king a land tax, effectively land rent, "which he assessed at his pleasure." They also owed the king their service, or labor, which he could call for at any time, along with "some proportion of the productions of the land." If a chief or tenant failed to pay any of these, he could forfeit his right to use the land. In addition, land was typically redistributed whenever a new monarch assumed power.[13]

In 1846 the Kingdom of Hawaiʻi's Legislative Council described the islands' system of land tenure as follows:

> When the islands were conquered by Kamehameha I, he followed the example of his predecessors, and divided out the lands among his principal warrior chiefs, retaining, however, a portion in his own hands to be cultivated or managed by his own immediate servants or attendants. Each principal chief divided his lands anew and gave them out to an inferior order of chiefs by whom they were subdivided again and again, often passing through the hands of four, five or six persons from the King down to the lowest class of tenants. All these persons were considered to have rights in the lands or the productions of them, the proportions of which rights were not clearly defined, although universally acknowledged.[14]

Because of its importance for kalo production, irrigation was carefully regulated. Irrigation water was diverted from streams through ditches to kalo and other agricultural fields. According to O'Shaughnessy, the ditches, called ʻauwai, were "excavated in the surface earth and repaired by the joint users, each of whom had to devote so many days towards maintenance." The irrigation water was divided among users by "set rules and at stated times, each district within the branch supply ditch getting so many hours' flow of the stream."[15] Construction of ʻauwai was organized by the chief of the ahupuaʻa, and where a ditch system served more than one ahupuaʻa, the amount of water allocated to each was proportional to the labor provided by each chief for its construction. H. A. Wadsworth speculated on how the correct slope of the ditch was likely determined. "It has been said that a tool, in the form of a long bamboo pole with the nodal tissue removed, was used for this purpose, since water poured into the upper end escaped at the proper rate when the slope was correct."[16]

The last part of the system to be built was the dam, "always a low loose wall of stones with a few clods here and there, high enough to raise water sufficiently to flow into the 'auwai." No more than half of the water in the stream was diverted into the 'auwai, and whenever a dam was rebuilt, "delegates from each dam below were required to be present to see that a due proportion of water was left in the stream."[17]

Though the design and building of ditches fell under control of the chief of the ahupua'a, day-to-day management was the responsibility of an agent, or *konohiki*, he appointed. This official controlled distribution of water to individual kalo patches, usually intermittently on the basis of the time needed to irrigate each plot. The konohiki could also call on water users for labor needed to maintain the 'auwai and withhold water from any who refused.[18]

Hawaiian water rights were assigned to lo'i, or plots of irrigated kalo. Plantings of sweet potatoes, sugarcane, and bananas generally did not receive water rights, though they were often planted on the berms surrounding the lo'i and received excess water when it was available.[19]

Mrs. Emma Metcalf Nakuina, Commissioner of Private Ways and Water Rights for the District of Kona, O'ahu, gives a good description of the feast that followed completion of a new 'auwai.

This was an occasion for rejoicing and feasting, and was never hurriedly done. The water kahuna or priest had first to be consulted in regard to a favorable day, which being settled, the konohiki was required to furnish a hog large enough to supply a good meal for all the workers of the 'auwai, red fish ('ahuluhulu), 'ama'ama and āholehole as well as 'awa root for the use of the priest at the opening ceremonies; . . . kukui nut and poi galore. On the appointed day, all the workers decked in leis [lei] . . . proceeded to the end of the 'auwai nearest the spot chosen for the dam, each one bearing a stick of firewood for the imu or oven in which the hog and other articles of food was to be baked. The imu was made in the 'auwai near the point where the water was to enter it; the hog, lū'au potatoes and kalo, or taro, were placed in it, and while these were cooking, the 'awa root was chewed or pounded and strained, and the fish lawalued (wrapped in ti leaves and roasted over coals).

When everything was cooked and in readiness, the water kahuna took the head of the hog, the fishes, and a bowl of 'awa juice, and going to the place where the dam was to be built made an offering of these to the water Akua or God. An invocation would be made and a petition that the local god or goddess would take the dam and 'auwai under his or her especial protection, not only sending or causing a good supply of water to fill the stream at all times, so that her votaries might be pleased with good and abundant crops, but also to guard against both drought and floods as being disastrous to the planting interests. . . . Everything at this feast of consecration had to be consumed either by the people or by their dogs. All the refuse was buried in the imu; the dam built in a few minutes, and the water turned in to the new 'auwai; flowing directly over the now submerged imu. The younger folks would likely indulge in bathing in the pool formed by the dam, while the older ones with the konohikis and invited guests would follow the water through the new-made 'auwai, and singers of both sexes would chant songs composed in honor of him who had planned and carried out the beneficent undertaking that would be the means of a supply of food for many.[20]

The quantity of water allocated to each rights holder was proportional to the amount of labor furnished for construction of the 'auwai. Water from the 'auwai was diverted to the fields based on time, which could be a few hours to more than a day per user. The time was determined by the movement of the sun during the day and the stars at night. When it was time for a user to begin taking water from the 'auwai, he went to the dam with the konohiki (or water master, *luna wai*) to assure that it was in proper condition. The user then followed the 'auwai to his fields, removing any debris from the ditch and shutting off water from any secondary 'auwai or fields that were not authorized to receive water at that time.[21]

Most ahupua'a contained a few small lo'i in their upper areas, generally on steep lands, which could not be irrigated like normal lowland taro plots. These lo'i were generally awarded *kula*, or small-volume, continuous irrigation waters. Mrs. Nakuina recalled, "In good seasons when there was plenty of water in summer, surplus water was sometimes led on to kula land and a second crop of [sweet] potatoes planted, but this was never done if any lo'i . . . should be needing the water."[22]

Hawaiian water laws were strictly enforced, and "any one in olden times caught breaking a dam built in accordance with the Hawaiians' idea of justice and equity would be slain by the share holders of that dam, and his body put in the breach he had made, as a temporary stopgap, thus serving as a warning to others who might be inclined to act similarly." However, excess water was shared when water was scarce, and the luna wai could "take water from parties . . . after their patches or loʻi were full but before their time had expired, and turn it on to any loʻi that was suffering whose turn in the rotation had not arrived."[23]

In the early 1800s, Archibald Campbell described the typical Hawaiian agriculture on the dry leeward side of Oʻahu west of Honolulu, an area that was similar on the leeward coastal areas of Maui and other islands. "The roads and numerous houses [were] shaded by cocoa-nut [coconut] trees, and the sides of the mountains covered with wood to a great height." The flat land along the shore was used to produce taro, yams, and sweet potatoes. The two latter crops were grown without irrigation, but where irrigation was possible, taro was "the chief object of their husbandry, being the principal article of food amongst every class of inhabitants." The system of taro cultivation was "extremely laborious, as it [was] necessary to have the whole field under water; it [was] raised in small patches, which [were] seldom above a hundred yards square; these [were] surrounded by embankments, generally about six feet high, the sides of which [were] planted with sugar canes, with a walk on top; the fields [were] intersected by drains or aqueducts, constructed with great labour and ingenuity, for the purpose of supplying the water necessary to cover them."

Taro (kalo) cultivation involved digging and leveling the soil with "a wooden spade, called *maiai,* which the labourers use[d], squatting on their hams and heels. After this it [was] firmly beat down, by treading on it with their feet till it [was] close enough to contain water." The taro was propagated "by planting a small cutting from the upper part of the root with the leaves adhering. The water was then let in, and cover[ed] the surface to a depth of twelve to eighteen inches; in about nine months they [were] ready for taking up." Campbell remarked that he had "often seen the king working hard in a taro patch," possibly to set "an example of industry to his subjects." However, this "could scarcely be thought necessary among these islanders, who [were] certainly the most industrious people" he had ever seen.[24]

Most of Campbell's experience with irrigation and taro production was near Pearl Harbor, where land was flat and abundant spring water was available, making taro patches with high levees and deep water feasible. However, citing Professor John Wise of the University of Hawaiʻi as his authority, H. A. Wadsworth stated:

Natural spring waters were utilized by surrounding the spring with high, water-tight levees, which enclosed about half an acre of ground. Piles of dirt thrown up inside these basins provided planting area for the crops to be grown. Bananas planted on the tops of the piles secured adequate moisture without submergence while sugar cane and yams planted at lower elevations, and of course, nearer the water level, supposedly secured moisture in accordance with their needs. Taro planted still lower on the slopes was partially submerged. This method of culture was necessarily limited to areas in which natural springs occurred under such conditions that a two- or three-foot head might be maintained.[25]

On more sloping lands, lower and narrower levees and shallower flooding were the norm, with "low levees thrown around conveniently shaped areas of land and water admitted from the neighboring 'auwai. Apparently water was admitted to each basin from the one above it, if not from the 'auwai itself, drainage from the last patch finding its way into the original stream or another ditch." By 1933 the best remnants of old taro patches were found in the Kalalau Valley of Kaua'i, where "in no instance were banks more than three feet high discovered, except in cases where topography demanded distinct terracing. Nor were the tops of the banks of significant width." Wadsworth also noted that "taro production in the early days was a continuous operation, harvesting and replanting being determined primarily by the rate of consumption by the kuleana holder and his dependents. Supplies of mature tubers (corms) were removed from the beds as needed and carried away from the growing area for cleaning and preparation as poi. Replacements of planting material in the beds provided a continuous supply." The native Hawaiian system of taro production was very productive, and Wadsworth cites reports that "one square mile planted to the same vegetable [taro] will feed fifteen thousand one hundred and fifty-one persons."[26]

According to Campbell, "When the islands were discovered, pigs and dogs were the only useful animals they possessed," but because of "preservation of the breeds left by Vancouver [in 1793] and the other navigators . . . in a short time the stock of horned cattle, horses, sheep, and goats will be abundant." The island of Hawai'i already had "many hundreds of cattle running wild," and on O'ahu, "there was a herd of nine or ten upon the north side of the island." Campbell added that "the cattle lately introduced [were] pastured upon the hills and those parts of the country not under cultivation, the fences not being sufficient to confine them. The hogs [were] kept in pens, and fed on taro leaves, sugar-canes, and garbage." Sheep and goats were "already very numerous," and the king "had five horses, of which he was very fond, and used frequently to go out on horseback."[27]

Substantial changes in vegetation, agriculture, and land tenure began to occur during the early 1800s due to the increasing influence of foreign visitors and immigrants. At the same time, the native Hawaiian population was in decline as a result of introduced diseases to which they had little immunity. As contact with industrialized countries grew, the need increased to generate foreign exchange to purchase manufactured goods. Sale of sandalwood, which was used in China for incense and fine wood products, began in 1790, and it was the chief source of wealth for the islands in the first quarter of the nineteenth century. It sold for $8–10 per picule of 133.5 pounds, but by 1825 the supply had been largely depleted. Harvesting the ever-rarer trees was prohibited in 1839, and seven years later the forests were declared property of the monarchy. The first whaling ships had arrived in Honolulu in 1820, and their numbers increased rapidly until over 100 arrived each year. Supplying provisions to the ships and recreation to their crews became a principal source of revenue for the Kingdom as the sandalwood trade declined.[28]

Several factors were responsible for the destruction of water-conserving forests on all of Hawai'i's islands. While the agricultural practices of the Hawaiians—including the use of fire to clear lands, the cutting of sandalwood for export to China, and the burning of other trees primarily for fuel by the sugar factories—certainly did not help matters, it was the introduction of cattle by Captain George Vancouver in 1793 that did the most damage. The cattle roamed freely in the Hawai'i forests owing to a kapu (prohibition) placed on their harm by Kamehameha I. Dr. William Hillebrand, a prominent medical doctor in Honolulu, had given a remarkable, visionary speech in 1856 before the Royal Hawaiian Agricultural Society, making a plea for the protection of the remaining forests from cattle grazing and the reforestation of those that had already suffered damage. "It cannot be denied," he wrote, "that in many places the domain of the forest has been seriously encroached upon by man, and more by cattle," which he considered the worst "of all the destroying influences man brings to bear on nature." Extolling the benefits of reforestation, Hillebrand contended, "Nowhere, I imagine, shall we have to wait more than six years before some results are obtained [and] . . . on the banks of rivers and in low, wet places our end may be attained in less than four years." He favored reforestation with both native and adapted introduced species, predicting that as soon as the hills were reforested, "springs will soon bubble forth again from every nook and corner, between rocks and from under the fern tree, to unite in streams which encased in a framework of bamboo canes and lianas will pour out of every valley and cover with fertility the now arid and sterile plains."[29] Hillebrand's views about the importance of forests became widespread, prompting legislation in 1876 to protect the forests on government lands.

Changes in Land Tenure and Water Rights

Many local chiefs who had previously lived within their ahupua'a moved to the towns to take advantage of foreign goods and services, appointing konohiki to represent them and manage the irrigation systems of the ahupua'a. This loosened the ties between the chiefs and their tenants and led to increased abuse of tenants by the konohiki. In response to weakening lines of authority, the traditional obligations of chiefs and tenants were codified in the Constitution of 1840. Thereafter, tenants were required to work 36 days per year for the king and an equal amount for the local chief.[30]

In addition, laws were passed in 1839 and 1840 to prevent eviction or seizure of property without cause, though these decrees soon proved to be "totally inadequate." In order to bring more order to claims of land rights, a Board of Commissioners to Quiet Land Titles was established in 1846 and charged with overseeing the process of filing claims, taking testimony, surveying boundaries, and awarding claimants land rights. These rights were not equivalent to fee-simple ownership; rather, awards were meant to serve as grounds to receive "a Royal patent in fee simple from the Minister of the Interior, on payment of a commutation to be agreed upon by His Majesty." The board fixed the cost of obtaining fee-simple title of a claim at one-third of the unimproved value of the land. The board spent the rest of 1846 and most of 1847 establishing claims of foreigners for land and house lots in the Honolulu area. Too much confusion reigned for the board to address division of lands among the king, local chiefs, and the government. As the islands' foreign population grew, so did calls for the monarchy to permit them to buy farmland to produce commodities both for export and for sale to the whaling and trading ships that plied Hawaiian waters. The pressure grew until the monarchy's absolute control of the land became untenable, and in December 1847 King Kamehameha III and chiefs in a privy council agreed to divide the Kingdom's lands as follows. One-third would be retained by the king as his private property. Of the remaining two-thirds, one-third would go to the Hawaiian government, one-third to the chiefs and their konohiki, and one-third to the tenants, "the actual possessors and cultivators of the soil, to have and to hold, to them, their heirs and successors forever." This division of lands was known as the "Māhele" or "Great Māhele," and between 1848 and 1850 the details of the divisions of land among the king, the chiefs, and the government were negotiated, concluded, and recorded in the "Māhele Book" or book of division.[31]

The Great Māhele permitted private ownership of land to individuals and corporations, as well as the possibility of leasing the right to divert water from government lands. The islands soon became "a country of farmers," in which professionals and skilled laborers "had farms on which they worked when not busy at their trades." The board of commissioners remained active until it was dissolved in 1855, by which time it had confirmed 11,309 claims. During the 1850s and 1860s additional acts were passed to award lands to certain konohiki who had failed to make their claims to the board prior to the original February 1848 deadline for presenting claims. Other acts clarified titles to the personal lands of the heirs of Kamehameha III, defined fishing rights, outlined boundaries of claims, and prescribed the sale of government lands at auction. Between 1850 and 1860 much of the desirable government land was sold, mostly to native Hawaiians who were willing to then sell to foreign residents. In 1884 the government offered public lands for settlement as homesteads. These lands were initially limited to 20 acres of dry, or kula, land or two acres of wet, or taro, land. The dry land limit was later raised to 100 acres for some of the rocky districts of Maui and Hawai'i. The applicant was allowed to occupy the land for five years free of taxes if he or she built a house within one year, fenced the property within two years, and paid interest on the note at 10 percent. At the end of five years the owner could pay the balance of the note in full or secure a mortgage for the unpaid balance. Since the quality of most original land surveys was poor, a great deal of work was needed to resolve gaps and overlaps between claims and to develop general maps of land ownership within districts.[32]

In 1860 an act was passed to resolve controversies regarding water rights and rights of way. Decisions were in the hands of three appointed commissioners in each election district, and appeals could be made to the Circuit and Supreme courts. In 1888 the three commissioners were reduced to one, and appeals were limited to the Supreme Court. Finally, in 1907 the office of commissioner was abolished and its duties were assumed by circuit judges. Throughout this period, proceedings were generally "simple, expeditious, inexpensive, and informal." In addition to settling numerous conflicts and defining precise rights to quantity and timing of irrigation, the process had established a number of general principles. For example, prescriptive rights (based on long periods of uncontested use) could be granted for larger quantities of water than the landowner originally had a right to use. Prescriptive rights could also be obtained to irrigate kula lands, to use the overflow and seepage from neighboring taro patches and streams, and to change the time of water use. Failure to use a water right did not automatically lead to forfeiture of the right, and as long as others' rights were not damaged, water could be diverted by the owner to other land, including other kuleana or ahupua'a, even if the land to which it was transferred had no water right. Finally, any surplus water in a stream was the property of the konohiki to allocate as he chose.[33]

Agriculture Booms

The California gold rush that began in 1849 strengthened markets for Hawaiian agricultural products considerably. In addition to sugar, it created a strong demand for farm labor in response to the new export markets for potatoes, sweet potatoes, onions, pumpkins, oranges, molasses, coffee, and hogs.[34]

> Between the years 1850 and 1860 a large part of the government land was sold to the common people in small tracts at nominal prices. The rapid settlement of California opened a new market for the productions of the Islands, and gave a great stimulus to agriculture. For a time large profits were made by raising potatoes for the California market. Wheat was cultivated in the Makawao district [of Maui], and a steam flouring mill was erected in Honolulu in 1854. The next year 463 barrels of Hawaiian flour were exported. A coffee plantation was started at Hanalei, Kaua'i, in 1842, and promised well, but was attacked by blight after the severe drought of 1851–2. The export of coffee rose to 208,000 pounds in 1850, but then fell off. The export of sugar only reached 500 tons in 1853. The sugar mills were generally worked by oxen or mules, and the molasses drained in the old fashioned way.[35]

The whaling industry declined precipitously during the 1860s due to the replacement of whale oil by kerosene produced by the fledgling oil industry in the United States. The number of whaling ships visiting Honolulu decreased from a peak of 549 in 1859, to 325 in 1860, 190 in 1861, and just 73 in 1862. As the number of whalers decreased, the economy of the islands suffered, and many of its leaders began to look to agriculture as a possible source of profits. The Royal Hawaiian Agricultural Society, which had been formed in 1850, purchased land on Oʻahu for the testing of plant introductions, and in 1858 and 1859 the society experimented unsuccessfully with rice production. But in 1860 a small amount of rice seed was obtained from South Carolina, and small plots grown by Dr. S. P. Ford in taro fields near Honolulu yielded abundantly. Word of Ford's success spread quickly, prompting a story in the *Honolulu Advertiser* that read: "Everybody and his wife (including defunct government employees) are into rice—sugar is nowhere and cotton is no longer king."

Rice fever led to a situation in which taro patches were bought and rented "at excessive prices," taro plants were pulled up and the land planted to rice, and high prices were paid for labor to plant and tend the rice paddies. As a result, taro became scarce, and its price rose beyond the means of many native Hawaiians. Because the capacity to mill and polish the rice was very limited, most of the 1862 crop was sent to California as rough rice (paddy). But milling capacity eventually increased; from 1865 to 1875 the industry not only provided about 2 million pounds of milled rice per year for consumption in the islands but also exported up to 2.1 million pounds of milled plus rough rice. This made it the second-most important crop produced in the islands.[36]

In 1874 Isabella Bird, a British world traveler, spent several months traversing the islands, observing the people and the land. Her letters to friends in England provide riveting accounts of riding on horseback to the top of the Kīlauea, the active volcano on the island of Hawaiʻi, and riding the steep and slippery trails along the islands' windward coasts to view the plight of declining native Hawaiian villages. Bird also reported on the political uncertainty and sense of optimism that prevailed as the Kingdom negotiated with the United States to eliminate the tariffs that were holding back development of the sugar industry. She even found time to admire the islands' flora and geology, and her descriptions of the islands' agriculture, including the Onomea Sugar Company near Hilo, are among the best we have from this critical period in the industry's history. Bird provided a succinct summary of diversified agriculture in the islands.

The value of domestic exports in 1873 was $1,725,507. Among these are bananas, pineapples, pulu [tree fern fiber], cocoanuts [coconut], oranges, limes, sandal-wood, tamarinds, betel leaves, shark's fins, paiai [pounded but undiluted taro corm], whale oil, sperm oil, cocoanut oil, and whalebone. Among other commodities there was exported, of coffee 262,000 lbs., of fungus [a mushroom called pepeao by the Hawaiians, commonly referred to as wood ear] 57,000 lbs., of pea nuts 58,000 lbs., of cotton 8,000 lb., of rice 941,000 lbs., of paddy (unhulled rice) 507,000 lbs., of hides 20,000 packages, of goat skins 66,000, of horns 13,000, and of tallow 609,000 lbs.

The group [of islands] has an area of about 4,000,000 acres, of which about 200,000 may be regarded as arable, and 150,000 as specially adapted for the culture of sugar-cane. Sugar, the great staple production, gives employment in its cultivation and manufacture to nearly 4,000 hands. Only a fifteenth part of the estimated arable area is under cultivation. Over 6,000 natives are returned as the possessors of Kuleanas [kuleana] or freeholds, but many of these are heavily mortgaged. Many of the larger lands are held on lease from the crown or chiefs, and there are difficulties attending the purchase of small properties.

Almost all the roots and fruits of the torrid and temperate zones can be grown upon the islands, and the banana, kalo [taro], yam [*Dioscorea alata*], sweet potato, cocoanut [coconut], breadfruit, arrowroot [pia in Hawaiian, genus *Tacca*], sugar-cane, strawberry, raspberry, whortleberry, and native apple [(mountain apple), *Eugenia sp.*], are said to be indigenous.[37]

Beginnings of the Sugar Industry

Sugarcane was brought to the Hawaiian Islands by its Polynesian discoverers. It is not surprising, therefore, that immigrants from countries with a tradition of sugarcane production took advantage of this introduced cane and began producing sugar and molasses. A merchant from China, presumably engaged in the sandalwood trade, was probably one of the first people to produce sugar in the islands. He set up a stone mill and boilers on Lāna'i in 1802 but apparently harvested only one crop. Sugar was also made in Honolulu by Don Francisco de Paula Marin of Spain in 1819, and an Italian named Lavinia made sugar in 1820 by employing native Hawaiians to pound it with stone pestles on large wooden poi boards, then boil the juice in a small copper kettle. By 1823–1824 the manufacture of sugar and molasses was common in the islands, probably largely for the production of rum, which the American sugar refiner George Rolph reported "was extensively carried

Figure 1.1. Stone sugar mill with oxen. Photo courtesy of HARC

out at that time." In 1825 John Wilkinson of England attempted to grow cane on a commercial scale, planting 100 acres in the Mānoa Valley of Oʻahu, but he died in 1827 and the project was abandoned.[38]

It was not until the late 1830s that a successful, sustainable sugar plantation was finally established in the islands—on Kauaʻi. Its success established a number of precedents that profoundly affected the structure of sugar companies throughout the islands.

Kōloa Sugar Company, Kauaʻi—1830s

In 1835 William Hooper, working for the Honolulu-based trader Ladd and Company, established what many consider to be the first commercially successful sugar plantation in the islands at Kōloa, on the southern coast of Kauaʻi. The location of the plantation may have been influenced by the fact that the word *Kōloa* was derived from *kō* (cane) and *loa* (great), and cane grew wild in the area. Ladd and Company had leased 980 acres of land from the monarchy, agreeing to produce cane, compensate Hawaiian laborers with satisfactory wages, and pay taxes on each employee. Hooper began clearing land for the plantation with 25 workers, promising each one wages of $2 per month. But within days of beginning their jobs, the laborers demanded payment on a daily basis; Hooper agreed on a daily wage of one real (12.5 cents), but he didn't pay it in cash.

He created his own form of payment—plantation script in units of 12.5, 25, and 50 cents. The handmade cardboard coupons could be exchanged for consumer goods shipped by Ladd and Company from Honolulu to the company store at Kōloa. Hooper felt that by employing native Hawaiians for wages, he was emancipating them from the traditional system in which they owed labor to the local chief. In less than a year Hooper's workforce had increased to 40 men, and within a year he had 25 acres under cultivation. He and his workers had also built a mill pond, a water-powered mill, a boiling house, a sugar house, blacksmith and carpenter shops, 20 houses for his workers, and a house for the superintendent. The normal workday was defined by the plantation bell, beginning at 6:00 a.m. and ending at sunset, with time off for breakfast and lunch. Fridays were set aside for the workers to cultivate their kalo plots, and Saturdays were used for working in their gardens, cooking, and trading all kinds of food and other goods at the plantation store and at shops set up for the day. Every third week, Hooper provided his workers with a barrel of fish. The Hawaiians appear to have been more than willing to give up their traditional lifestyle for what the plantation provided: houses, food, and regular wages that could be used to purchase consumer goods at the plantation store.[39]

In just over a year Hooper's new water-powered wooden mill was up and running, requiring four men to feed it with cane. Shortly thereafter, in late 1837, an iron mill was acquired. But Hooper constantly complained to Ladd and Company that his workers were inefficient, lacked a proper work ethic, and required close and constant supervision. He also had trouble with shoplifting at the plantation store and counterfeiting of his script. In response, he hauled the shoplifters before the native authorities at Kōloa, who assigned them to work on the roads. He also had Ladd and Company order printed plantation script that would be more difficult to counterfeit.

Dissatisfied with his male workers, by 1838 Hooper had begun to hire Hawaiian women for 6 cents per day. He also brought on single Chinese men who were usually assigned to work in the mill, and he constructed barracks to house them. Within a year Ladd and Company was sending rice to the plantation for its Chinese workers. Exhausted by his efforts to establish and run the plantation, Hooper left Kōloa in 1839, having shipped 30 tons of sugar and 170 barrels of molasses to Honolulu in 1837 and 1838. More important, however, he had established a paternalistic model for later plantations, with the company providing food, wages, separate housing based on ethnicity, a plantation store, and a regular workday governed by the plantation bell.[40]

The success of the Kōloa Sugar Company was not lost on the islands' entrepreneurs, and other plantations were quickly established. William Ladd of Ladd and Company wrote in April 1838 of Hooper's efforts:

> It is a very common opinion that sugar will become a leading article of export. That this will become a sugar country is quite evident, if we may judge from the varieties of sugar cane now existing here, its adaptation to the soil, price of labor and a ready market. From experiments hitherto made, it is believed that sugar of a superior quality may be produced here. It may not be amiss to state that there are now in operation, or soon to be erected, twenty mills for crushing cane propelled by animal power, and two by water power.[41]

After Hooper left, the Kōloa Plantation suffered severe setbacks. By 1840 "two mill sites were abandoned, and the entire works, including buildings, machinery, and furnaces were sacrificed." But in 1841 a third mill was erected with new machinery, enabling the plantation to increase production, "though the quality of sugar at, and up to this time was very inferior." That quality improved greatly, however, after M. Prevost, who had "considerable experience in sugar manufacture in the Isle of Bourbon [Reunion]," was hired.[42] The plantation in Kōloa continued to increase its production; in 1850 it turned out $20,000 worth of sugar, with expenses less than half that amount. By 1857 it had 100 native Hawaiian and 20 Chinese field laborers and was producing 200 tons of sugar per year.[43]

Industry Expansion

The early expansion of the industry produced some false starts but also some successful enterprises. For example, Governor Huakini of the island of Hawai'i planted a few fields of cane at North Kohala, and in 1841 he entered into an agreement with a Chinese man known as Aiko to install a mill. The simple mill consisted of upright wooden rollers about eighteen inches in diameter and two and a half feet tall, bound with iron and powered by an overshot waterwheel.

The sugar he produced was "laboriously carted over the hills to Māhukona" where it was shipped to Honolulu. The carts were very heavy and the wheels were nothing more than cross sections of koa logs. As a result, the wear and tear on the oxen and carts was substantial. Nevertheless, Aiko was making money and would have remained, "but the heir of Governor Huakini so increased the rents of lands and other charges that he threw the whole thing up in disgust and left for Hilo, where he carried on a plantation with success for many years."[44]

Despite these difficulties, between 1836 and 1841 exports were impressive, with sugar yielding $36,000 and molasses bringing in $17,130. Production increased from about 40 tons in 1838 to 100–300 tons between 1844 and 1849, and a whopping 816 tons in 1859. Still, production was very erratic due to periodic droughts and plantation failures.[45]

Though early cane producers were slowly learning to grow the crop, they were not yet producing a good and consistent product, and prior to 1843 the quality of Hawaiian sugar "did not enjoy a favorable reputation abroad." Sugar was "shipped imperfectly cured," and in some cases was adulterated with "refuse . . . collected from the drying-houses." Not until after 1851—when the Royal Hawaiian Agricultural Society began to influence the situation and new technology became available to separate molasses from the sugar—was Hawai'i's reputation for "good merchantable sugars" established.[46]

Prior to 1850 most mills were crude wooden affairs. The cane was fed one or a few stalks at a time into the wooden rollers, which were powered by oxen, horses, or, rarely, water. The juice was boiled to thick syrup in whalers' trypots, large kettles used on whaling ships to render whale oil. It was allowed to cool and crystallize, "after which the granulated mass was packed into mats, bags, boxes or barrels with perforated bottoms for the molasses to drain off."[47] Fertilizer was not used, irrigation was not practiced, less than 50 percent of the sugar in the cane was extracted (compared with over 90 percent by the early twentieth century), and sugar yields were usually about one ton per acre. However, several technical innovations occurred in the 1850s, setting the stage for rapid industry expansion when sugar prices rose during the American Civil War. In 1851 D. M. Weston, manager of what later became Honolulu Iron Works, invented the first centrifugal sugar-drying machine to extract the molasses, greatly increasing the efficiency and quality of sugar production. Installed on the East Maui Plantation, the machine eliminated "the old and tedious process of draining from boxes or barrels, through holes bored to allow the molasses to drain off, leaving it but imperfectly or unevenly dried after all."[48]

Other innovations followed quickly. In 1856 the Lihu'e Sugar Plantation on Kaua'i developed the first extensive irrigation system in Hawai'i. Costing $7,000, it consisted of a 10-mile-long irrigation ditch and tunnel system. Beginning in 1858 steam engines were replacing animal traction to power new iron mills, which had replaced the much-less-efficient wooden mills.[49]

Average New York sugar prices rose from 4.4 cents per pound in the late 1840s to 6.95 cents in the 1850s. Kerosene began to replace whale oil for lamps, and capital began to move out of whaling and into sugar. As a result, between 1857 and 1861 the number of plantations on the Hawaiian Islands increased to 22, with nine powered by steam, 12 by water, and only one by animal traction. Production had grown to 722 tons of sugar and 2,600 barrels of molasses by 1860. Equally as important, the Civil War disrupted the supply of sugar from the southern United States, and New York sugar prices rose to nearly 10 cents between 1862 and 1869, increasing demand for Hawaiian sugar.

The increased demand led to the establishment of more sugarcane plantations, among them Makee and ʻHaikū—both on Maui. The ʻHaikū Plantation, which began operations in 1861, used steam for grinding, boiling, and drying its sugar. Vacuum pans to evaporate water from the juice were installed in mills around that time, first on Kaupakuea Plantation north of Hilo and the next year in Lahaina on West Maui. As a result of these technical advances, the efficiency with which sugar was extracted from juice increased dramatically. Consider the statistics over the decade and a half that began in 1850: That year only 5.8 pounds of sugar per gallon of molasses was suitable for export. This ratio increased to 7.6 pounds in 1855, 13.3 pounds in 1860, and an impressive 28.2 pounds per gallon in 1865. This, in addition to increased acreage in cane, allowed sugar exports to expand more than tenfold between 1860 and 1865 (see Table 1.1) and led to the establishment of a sugar refinery in Honolulu in 1861. However, "after a struggle of six years or so it closed its doors, never having done more than manufacture sugar from the molasses from the plantations and drain the pockets of its stock-holders."[50]

The growth of sugar exports during the 1850s and early 1860s was remarkable, but the 1870s saw a decline in New York's sugar prices to about 6 cents per pound; consequently, the rapid wartime rise in cane acreage and sugar production dropped off. Another problem faced by the industry was the 40 percent U.S. tariff on foreign sugar, which greatly reduced profits for Hawaiian producers. Because of the close ties of Hawaiʻi's missionary community to New England and the United States, and in an effort to promote both the sugar industry and the islands' economy, the Kingdom entered into negotiations with the United States for a reciprocity treaty to eliminate tariffs between the two countries. Negotiations failed to gain U.S. Senate approval in the 1850s and 1860s, despite two key factors: the Kingdom's role in supplying the California gold rush market, and the ongoing shortage of sugar in the northern United States during the Civil War.[51]

Table 1.1. Hawaiian sugar and molasses exports, 1837–1874

	Sugar	Molasses
Year	*Pounds*	*Gallons*
1837	4,286	2,700
1838	88,543	11,500
1839	100,000	75,000
1840	360,000	31,739
1841	60,000	6,000
1842–43	1,145,010	64,320
1844	513,684	27,026
1845	302,114	19,353
1846	300,000	16,000
1847	594,816	17,928
1848	499,533	28,978
1849	653,820	41,235
1850	750,238	129,432
1851	21,030	43,742
1852	699,170	62,030
1853	642,746	75,760
1854	575,777	68,372
1855	289,908	38,304
1856	554,805	58,842
1857	700,556	48,486
1858	1,204,061	75,181
1859	1,826,620	87,513
1860	1,444,271	108,613
1861	2,562,498	128,259
1862	3,005,603	130,445
1863	5,292,121	114,413
1864	10,414,441	340,436
1865	15,318,097	542,819
1866	17,729,161	851,795
1867	17,127,187	544,994
1868	18,312,926	492,839
1869	18,302,110	338,311
1870	18,783,639	216,662
1871	21,760,773	271,291
1872	16,995,402	192,105
1873	23,129,101	146,459
1874	24,566,611	90,000

Source: Thrum, *Almanac,* 1874.

Onomea Sugar Company, Hawai'i—Early 1870s

Isabella Bird's 1874 description of Onomea Sugar Company, located a few miles north of Hilo, highlights the degree to which plantation technology had advanced by the early 1870s.

This is a very busy season, and as this is a large plantation there is an appearance of great animation. There are five or six saddled horses usually tethered below the [plantation manager's] house; and with overseers, white and colored, and natives riding at full gallop, and people coming on all sorts of errands, the hum of the crushing-mill, the rush of water in the flumes, and the grind of the wagons carrying cane, there is no end of stir.

The plantations of the Hilo district enjoy special advantages, for by turning some of the innumerable streams into flumes the owners can bring a great part of their cane and all their wood for fuel down to the mills without other expense than the original cost of the woodwork.

The cane, stripped of its leaves, passes from the flumes under the rollers of the crushing-mill, where it is subjected to a pressure of five or six tons. One hundred pounds of cane under this process yield up from sixty-five to seventy-five pounds of juice. This juice passes, as a pale green cataract, into a trough, which conducts it into a vat, where it is dosed with quicklime to neutralize its acid, and is then run off into large vessels. At this stage the smell is abominable, and the turbid fluid, with a thick scum upon it, is simply disgusting. After a preliminary heating and skimming it is passed off into iron pans, several in a row, and boiled and skimmed, and ladled from one to the other till it reaches the last, which is nearest to the fire, and there it boils with the greatest violence, seething and foaming, bringing all the remaining scum to the surface. After the concentration has proceeded far enough, the action of the heat is suspended, and the reddish-brown, oily-looking liquid is drawn into the vacuum-pan till it is about a third full; the concentration is completed by boiling the juice in vacuo at a temperature of 150° [F], or even lower. As the boiling proceeds, the sugar boiler tests the contents of the pan by withdrawing a few drops, and holding them up to the light on his finger; and, by certain minute changes in their condition, he judges when it is time to add an additional quantity. When the pan is full, the contents have thickened into the consistency of thick gruel by the formation of minute crystals, and are then allowed to descend into a heater, where they are kept warm till they can be run into "forms" or tanks, where they are allowed to granulate. The liquid, or molasses, which remains after the first crystallization is returned to the vacuum pan and reboiled, and this reboiling of the drainings is repeated two or three times, with a gradually decreasing result in the quality and quantity of the sugar. The last process, which is used for getting rid of the treacle (uncrystallized syrup), is the most beautiful one. The mass of sugar and treacle is put into what are called "centrifugal pans," which are drums about three feet in diameter and two feet high, which make about 1,000 revolutions a minute. These have false interiors of wire gauze, and the mass is forced violently against their sides by centrifugal action, and they let the treacle whirl through, and retain the sugar crystals, which lie in a dry heap in the centre.

The cane is being flumed in with great rapidity, and the factory is working till late at night. The cane (fiber) from which the juice has been expressed . . . is dried and used as fuel for the furnace which supplies the steam power. The sugar is packed in kegs, and the cooper and carpenter, as well as other mechanics, are employed.

Mr. A.[Austin, the plantation manager] has 100 mules, but the greater part of their work is ploughing and hauling the kegs of sugar down to the cove, where in favorable weather they are put on board of a schooner for Honolulu.[52]

Sugarcane Varieties

Most of the earliest sugar companies harvested (or purchased) cane and made sugar from the noble canes (*Saccharum officinarum* varieties) that had been grown by Hawaiians for centuries. Only after plantations began to cultivate their own fields did they become concerned about planting more productive varieties. Thomas Thrum's *Hawaiian Almanac and Annual for 1875* described several varieties of sugarcane that were being grown in the islands after the Civil War—white, ribbon, purple, and arrowless cane. There were two types of "white canes." One was pale greenish-yellow, becoming more yellow as it aged. It had a rather soft rind, long-joints or internodes, and its buds were partially sunken in the stalk. It also produced numerous seed heads, also known as "tassels" or "arrows," during the fall of the year, and after harvest new stalks regrew vigorously from the plants' underground buds

Figure 1.2. Fluming cane from field to mill. Photo courtesy of HARC

(meaning the variety "ratooned" well). Another "white cane" was the color of straw when ripe, but it had inferior juice quality and was rarely cultivated. Thrum described "ribbon cane" or "striped cane" as green and purple, rich and juicy, and adapted to high elevations. There were two types of "purple canes," but they were not readily distinguished, their rinds and joints were hard, and they were difficult to grind. "Arrowless cane," also known as *pua 'ole*, could be harvested late since it did not flower.[53]

An important step was taken in 1854, when a Captain Edwards of the whaleship *George Washington* brought the Cuban and Lahaina cane varieties to Lahaina from Tahiti. George M. Chase, the U.S. Consul in Lahaina, planted the varieties in his garden. The two varieties were easily distinguishable because Lahaina had "long straight leaves of light color . . . covered with prickles at the base" while Cuban had "leaves of darker green, bending down in graceful curves, with no prickles." Cuban was initially the favorite cane "during the years 1861–2, when sugar making commenced in Lahaina," but soon thereafter the Lahaina variety, also known as Tahiti and *kō Pākē* (meaning Chinese cane), became the favorite. Lahaina was popular because of its rapid growth, deep root system, drought resistance, hard rind that defied rat damage, rich juice, and "compact, firm fibre" that enhanced its value as fuel. Planters eagerly began to plant the Lahaina variety, and it would remain the industry's standard into the early twentieth century.[54]

Labor

Hawaiians were the overwhelming source of agricultural labor in the early decades of the industry. However, many were reluctant to move away from the coastal areas to new sugar plantations that lacked housing and other facilities needed by their wives and children. In addition and perhaps even more important, Hawaiians, like natives of the Americas, had been isolated for many generations before Captain Cook's crew and subsequent visitors exposed them to deadly diseases like chickenpox, measles, cholera, smallpox, tuberculosis, and various venereal diseases. These foreign diseases had a devastating effect on native populations. Though we do not have exact census numbers for the late eighteenth century, the Hawaiian population began to decline after foreigners made contact, from roughly 130,000 in 1832 to about 70,000 in 1853, about 49,000 in 1875, and just 34,000 in 1890.[55]

Faced with declining numbers of laborers, the sugar plantations hired agents to employ a system similar to that used to recruit sailors. The agents offered cash advances to laborers who would sign contracts to work for a plantation. Agents also paid the fines of convicted criminals in return for contract labor. However, labor remained in short supply despite these tactics, and "in 1852 a small shipment of Chinese laborers was sent for."[56]

In 1859 ten South Sea Islanders arrived at Kaua'i and went to work at Kōloa Sugar Company. In 1868 the Kingdom of Hawai'i created the Bureau of Immigration, and a decade later it was instrumental in bringing 86 Micronesians to the islands. But islanders from other parts of the Pacific were undergoing the same declining populations as the Hawaiians, and it was soon apparent that this was not a rich source of labor for the industry.[57]

Seeking other sources of plantation labor, the Hawaiian Bureau of Immigration asked the Hawaiian consul in Japan to look into recruiting Japanese laborers for the sugar industry. The first group of 148 Japanese arrived in the late 1860s, and the flow of both Japanese and Chinese laborers would swell in later years.[58]

Laborers were brought to the islands under the Masters and Servants Act of 1850, which provided for fines and/or imprisonment for violation of labor contracts. However, the system was inefficient and unsatisfactory to both the plantation owners and the laborers. For example, free laborers could command higher wages than contract laborers, and good contract workers earned no more than their lazy or inefficient peers. In addition, employers could not discharge poor contract laborers without expensive legal action.[59]

In 1874 Isabella Bird provided this outsider's commentary on the labor situation at Onomea Sugar Company near Hilo:

> This plantation employs 185 hands, native and Chinese, and turns out 600 tons of sugar a year. The natives are much liked as labourers, being docile and on the whole willing; but native labour is hard to get, as the natives do not like to work for a term unless obliged, and a pernicious system of "advances" is practiced. The labourers hire themselves to the planters, in the case of natives usually for a year, by a contract which has to be signed by a notary public. The wages are about eight dollars a month with food, or eleven dollars without food, and the planters supply houses and medical attendance. The Chinese are imported as coolies, and usually contract to work for five years. As a matter of policy no less than for humanity the "hands" are well treated; for if a single instance of injustice were perpetrated on a plantation the factory might stand still the next year, for hardly a native would contract to serve again.
>
> The Chinese are quiet and industrious, but smoke opium, and are much addicted to gambling. Many of them save money, and, when their turn of service is over, set up stores, or grow vegetables for money. . . . Great tact, firmness, and knowledge of human nature are required in the manager of a plantation. The natives are at times disposed to shirk work without sufficient cause; the native lunas, or overseers, are not always reasonable, the Chinamen and natives do not always agree, and quarrels and entanglements arise, and everything is referred to the decision of the manager, who besides all things else, must know the exact amount of work which ought to be performed, both in the fields and factory, and see that it is done.[60]

Plantations on Maui

Though we have few detailed descriptions of the sugar plantations on Maui prior to the 1850s, beginning in the 1820s entrepreneurs were busy seeking their fortunes in sugar. Cane production began early, and H. A. Wadsworth noted that in 1823 an Italian named Antone Catalina and a man from China, Hungtai, established mills near Wailuku.[61] By 1828 a Portuguese immigrant, Antonio Silva, had a mill at Waikapū. That same year, two Chinese brothers—known as Atai and Ahung—established the Hungtai Company, a retail store in Honolulu. Around the same time, Atai set up a plantation and mill at Wailuku. As of 1841 the 150-acre plantation had announced its expansion to 250 acres and was producing both white sugar and white syrup for sale in the brothers' store. The quality of their sugar, said to be "not inferior to the best imported Loaf Sugar," suggests that their sugar processing included some type of clarification to remove colored contaminants.[62]

Fields of white cane were growing on the hills above Hāna in 1837, and during the 1830s, 1840s, and 1850s at least 11 mills and plantations operated on the island, though some lasted for only a short time. Between 1837 and 1850 David Malo, a Hawaiian with ties to the monarchy, grew cane at Lahaina, ground it with a crude mill, and boiled the juice for molasses in whalers' trypots. In 1843 a man from China named Ahpong owned a sugar mill in Lahaina and sugarcane acreage on the island of Hawaiʻi.[63]

Commercial sugar plantations on Maui rose and fell with regularity in the mid-nineteenth century. It was largely an experimental industry, and the entrepreneurs who ventured into it knew they were taking a huge financial risk. In 1850 Dr. Robert Wood and Ambrose Spencer started the East Maui Plantation near the town of Makawao. Two years later, sugar grower L. L. Torbert began managing a plantation at ʻUlupalakua on the dry southwestern slopes of East Maui. Between 1849 and 1855 Stephen Reynolds, a well-known Honolulu resident and store owner, ran the Hāliʻimaile Plantation on East Maui about midway between Pāʻia and Makawao. Charles Brewer and Captain James Makee bought Hāliʻimaile and its mule-powered mill in 1856 and hired Torbert, who had just sold his plantation at ʻUlupalakua to Makee, to manage it. Within three years, Brewer became sole owner of the plantation, subsequently known as the Brewer Plantation, and produced 200 tons of sugar per year until his death in 1863, when it was sold to Dr. Gerrit Judd of Judd, Wilder, and Judd.[64]

In Volume 1 of *The Planters' Monthly* (1883) George Wilfong related some of his early experience with sugar and molasses production on Maui. He recalled that in 1849 Judge Alfred Parsons had "a sugar mill erected there [in Lahaina] for the purpose of making syrup, which was disposed of to whaleships at a good price." Parsons's sugar works was "composed of one wooden mill and three try pots [trypots] bought from whaleships . . . set on stones and adobe, no lime or cement being used in setting them." John Gower quickly bought Parsons's sugar works and moved it to Makawao, where he had purchased the McLane Plantation "which was one of the best estates of that day." In 1850 A. B. Howe bought the Hāna Plantation on the eastern shore of Maui from a Mr. Lingrin, "who had planted 60 acres of cane and put up a primitive sugar works in grass houses." The works consisted of "four try pots [trypots] which had been bought from whaleships, and a wooden mill worked by four yoke of oxen." Wilfong took over as manager of the Hāna Plantation for Howe in 1852, acquired additional land, and began "grinding wild cane, which was growing all about the land everywhere." Two kinds of wild cane were available, "the old white cane which grew very high and straight, and a smaller kind of cane, [which] the natives said . . . came from Tahiti. It grew wild, very straight, and the leaves dropped off as the cane matured." The latter was probably the aforementioned Lahaina variety, and Wilfong also planted both kinds of cane, "which grew nicely by cultivation, and brought forth a good crop." When his sugar boiler died, Wilfong—with the help of "two men who had worked with the sugar boiler"—taught himself the art of making strikes, or transferring the concentrated cane juice, at just the right consistency, to coolers where it would crystallize.[65]

Howe sent Wilfong five cast iron sugar kettles and an iron mill. Wilfong, his blacksmith, and a mason installed the "train" of five open kettles in a long masonry base under which a fire burned to boil the juice in the kettles. The fire was managed so that it was hottest under the last kettle, which contained the most concentrated syrup. Cane juice was ladled into the first kettle where it boiled and the sugar became more concentrated. It was then passed to the second kettle, which, because the juice contained less water, boiled at a slightly higher temperature. As the increasingly concentrated syrup passed to each successive kettle in the train, it boiled at a higher temperature until it finally reached the fifth kettle, in which the syrup was concentrated enough to crystallize when cooled. Wilfong recalled:

[At] about this time one of [iron works manager] Mr. Weston's centrifugal machines [used for separating the molasses from the sugar crystals] was sent up, a four man power attached, by wheel and belt. This was one of the first Mr. Weston ever made. I had it set according to directions, with a post in the ground eight feet deep, filled in with stones and cement; the spindle of the machine was inserted in the end of the post, on which the machine ran, and the molasses led off to the tank below. This machine was a curiosity to the people, and sometimes a hundred or more natives came to see it run. The school boys would come every afternoon and run the machine until night without pay, and would dispute their turn to run it. They would make it hum.[66]

The Hāna Plantation was purchased by Judd, but Wilfong continued to manage it. He constructed a new sugar house and produced cane on 150 acres. Wilfong made 60 or 70 tons of sugar a year and had no problem selling his molasses to whalers. But before he could ship his sugar, the buildings where he boiled his juice and crystallized his sugar burned, destroying over 50 tons of sugar, as well as the molasses tank and its contents. Unable to get credit to rebuild—several other plantations had recently failed—he "turned over the Hāna plantation to Dr. Judd, everything as it stood."[67]

Despite this setback, Wilfong's experience of over 30 years taught him that "the way to make good sugar . . . is to commence the work in the field. Plant good seed and at the right time, according to the elevation of your land, cultivate well, strip [the dead leaves from the stalk] and keep the cane clean and clear of rattoons [young secondary shoots]; then there will be no difficulty in making sugar of good quality. It only requires a careful hand in the sugar house, and men that understand machinery and the manufacture."[68]

Dorrance's excellent history of the Hawai'i sugar industry, *Sugar Islands,* identifies 12 commercially viable sugar plantations that operated on Maui between 1850 and 1875, beginning in 1850 with the East Maui Plantation Company. On Maui, as on the other islands, plantations were usually established during times of great demand and high sugar prices. For example, 10 new Maui plantations began operations in the 1850s and early 1860s in response to high sugar prices resulting from the California gold rush and the blockade of Confederate ports during the American Civil War. Nine of the 12 plantations established between 1850 and 1875 were located on the dry western slopes of East Maui, the isthmus between the two volcanoes, or the eastern slopes of West Maui. Only three of these 12 (Hāna Plantation Company, Wailuku Sugar Company, and Pioneer Mill Company) survived until 1900; most failed or were taken over by larger plantations in the 1880s and 1890s (Table 1.2).[69]

The advances in sugar milling and production technology during the 1850s, 1860s, and 1870s were substantial. In 1851 one of the new centrifuges was installed on the East Maui Plantation to speed separation of the sugar crystals from the syrup and molasses. Vacuum pans were introduced in the 1860s to lower the temperature needed to boil the juice and avoid scorching it as it was being concentrated. By the 1870s the Wailuku mill was being powered by a large waterwheel using water diverted from streams on West Maui, and early photos show a complex of flumes that diverted water from West Maui streams to float cane to the mill.[70]

Table 1.2. Plantations established on Maui prior to
1876, with dates of operation

*On the dry, leeward western and northwestern slopes
of East Maui*

East Maui Plantation Company	1850–1886
'Haikū Sugar Company	1858–1883
'Ulupalakua (Makee) Plantation	1858–1883
Grove Ranch, Plantation	1863–1889
Alexander & Baldwin	1872–1883

On the wet, windward side of East Maui

Hāna Plantation Company	1850–1904

On the eastern side of West Maui:

Waiheʻe Sugar Company	1862–1894
Wailuku Sugar Company	1862–1988
Waikapu Sugar Company	1862–1894
Bal and Adams	1865–1877

On the dry, leeward western side of West Maui

Pioneer Mill Company	1863–1999
West Maui Plantation	1871–1878

Source: Dorrance, *Sugar Islands,* 2000.

Lahaina means "cruel sun" in Hawaiian, appropriate to its location on the hot and dry leeward western side of West Maui. From the 1820s to the 1860s it was a thriving port for whalers, providing supplies, rest, and recreation for crews from across the Pacific. In 1863 Pioneer Mill Company was built by James Campbell and Henry Turton on cane land owned by Benjamin Pitman, a Hilo merchant. Beginning with a steam-powered mill and 126 acres, the mill ground cane from its own lands as well as cane supplied by independent growers, including Dwight Baldwin and his brother Henry, who eventually became, along with Samuel Alexander, founders of Alexander & Baldwin on East Maui. By 1866 Pioneer Mill was producing 500 tons of sugar a year, increasing its output to 1,000 tons in 1872. Pioneer Mill's growth in later years would be the basis for most of the sugar production on leeward West Maui, and eventually the tourist industry in the area.[71]

In 1874 Isabella Bird described Lahaina, the Pioneer Mill fields, and the mountains behind as "a picturesque collection of low, one-storied, thatched houses, many of frame, painted white; others of grass, but all with deep, cool verandahs, half hidden among palms, bananas, kukuis, breadfruit, and mangoes, dark groves against gentle slopes behind, covered with sugar-cane of a bright pea-green . . . between the ocean and the red, flaring, almost

inaccessible Maui hills . . . pinnacled, chasmed, buttressed, and almost ver-
dureless, except in a few deep clefts, green and cool with ferns and candlenut trees [kukui], and moist with falling water."[72]

'Ulupalakua (Makee) Plantation, situated high on the dry southwestern
slopes of East Maui, was purchased in the late 1850s by James Makee. The
cool temperatures at this altitude and lack of adequate rainfall or irrigation
caused the cane to grow slowly and kept yields low. But in 1867 its steam-
powered mill produced about 800 tons of sugar from 800 acres harvested.[73]
Isabella Bird described the plantation, located at an elevation of about 2,000
feet, as "one of the finest on the island." But water was very scarce, and "all
that is used in the boiling-house and elsewhere has been carefully led into
concrete tanks for storage, and even the walks in the proprietor's beautiful
garden are laid with cement for the same purpose." In addition, Makee had
planted "many thousand Australian eucalyptus trees on the hillside in the
hope of procuring a larger rainfall."[74]

Responding to increasing sugar prices during the Civil War, Wailuku Sugar
Company was formed in 1862 on 500 acres at the present town of Wailuku
on the east side of West Maui. That same year, Waihe'e Sugar Company was
formed on the northeast coast of West Maui north of Wailuku, and soon
had 800 acres of cane under cultivation. Also in 1862 the Waikapū Sugar
Company was organized to the south of Wailuku. Wailuku Sugar Company
was taken over by C. Brewer & Company in 1864 and incorporated in 1875.
It later absorbed Waihe'e and Waikapū Sugar Companies, as well as Bal &
Adams, formed in 1865, and Bailey Brothers, organized in 1884. By the
early twentieth century Wailuku was Maui's second-largest plantation.[75]

Samuel T. Alexander and Henry P. Baldwin were boyhood friends and
the sons of missionaries who in 1863 found themselves on Maui. Alexander was manager of the Waihe'e mill, at that time the largest of the three
Central Maui mills, and Baldwin was working for his brother, who was raising sugarcane in Lahaina. Alexander offered Baldwin a job as his assistant
at the Waihe'e mill, beginning what was probably the most important relationship in the history of the sugar industry on Maui. In 1869 Alexander
and Baldwin, while still working at Waihe'e, bought 12 acres of land near
Makawao for $110 to grow cane. The next year they purchased an addi-
tional 559 acres for $8,000 from a Captain Bush, beginning acquisition of
a plantation that would be known as Alexander & Baldwin Plantation and
eventually Pā'ia Plantation. In 1870 the partners entered into an agreement
with Robert Hind to build a mill at Paliuli to grind their cane, agreeing to
furnish enough cane to make 200 tons of sugar from the 1870–1871 crop

and 500 tons per year for the next nine years. Baldwin managed the Alexander & Baldwin Plantation while Alexander managed the neighboring 'Haikū Plantation. In 1872 the partners bought out Hind's interest in the mill. Also in 1872, Claus Spreckels founded Hawaiian Commercial Company and built a modern mill at Spreckelsville. The company, which would later become Hawaiian Commercial & Sugar Company, became a major competitor to Alexander & Baldwin, as well as to other Maui companies.[76] Thus, by the mid-1870s Maui's sugar entrepreneurs and plantations were positioned to take advantage of the next two major events that propelled the industry's rapid growth during the last quarter of the nineteenth century—the Kingdom of Hawaii's Reciprocity Treaty with the United States and development of major irrigation systems on Maui. The complicated pattern of plantation startups, failures, and mergers will be addressed in the next chapter.

Politics and Sugar

Problems faced by the industry in the early 1870s included a chronic lack of labor, depressed sugar prices following the Civil War, interest rates of 10–12 percent, and agents' commissions of 5–19 percent for selling their sugar in the United States. The high costs of producing and exporting sugar to the States did not abate, and the industry continued to lobby for a treaty between the Kingdom of Hawai'i and the United States to reduce tariffs on trade. But as long as the U.S. tariff remained in force, sugar planting in Hawai'i was at best a risky business.[77]

But optimism was returning with increased speculation about possible annexation of Hawai'i by the United States or a treaty between the two nations to relieve the heavy duties on sugar. Isabella Bird noted that "sugar is now the great interest of the islands. Christian missions and whaling have had their day, and now people talk of sugar. Hawai'i thrills to the news of a cent up or a cent down in the American market." If either reciprocity or annexation could be achieved, much more of the "200,000 acres productive soil on the islands, of which only a fifteenth is under cultivation" could be used to produce sugarcane and other crops.[78]

Those in the sugar industry were right to be excited. Hawai'i had many advantages over sugar-producing areas in the southern United States. Bird observed that "the magnificent climate makes it a very easy crop to grow," and "were labour plentiful and duties removed, fortunes might be made . . . Two and a half tons to the acre is a common [sugar] yield, five tons a frequent one, and instances are known of the slowly matured cane of a high altitude yielding as much as seven tons!"[79] The ability, due to lack of frost, to plant and harvest throughout the year was another advantage enjoyed by plantations in the islands.

> There is no brief harvest time with its rush, hurry, and frantic demand for labour, nor frost to render necessary the hasty cutting of an immature crop. The same number of hands is kept on all the year round. The planters can plant pretty much when they please, or not plant at all, for two or three years, the only difference in the latter case being that the rattoons [ratoons] which spring up after the cutting of the former crop are smaller in bulk. They can cut when they please, whether the cane be tasseled or not, and they can plant, cut, and grind at one time!
>
> Herein is a prospective Utopia, and people are always dreaming of the sugar-growing capacities of the belt of rich disintegrated lava which slopes upwards from the sea to the bases of the mountains.[80]

Bird gives us an outsider's view of the forces pushing for annexation or reciprocity in 1874, including the reaction of native Hawaiians to their proposals.

> There is a ferment going on in this kingdom, mainly got up by the sugar planters and the interests dependent on them, and two political lectures have lately been given in the large hall of the hotel in advocacy of their views; one, on annexation, by Mr. Phillips . . . and the other, on a reciprocity treaty, by Mr. Carter. Both were crowded by ladies and gentlemen, and the first was most enthusiastically received.
>
> Sugar is the reigning interest on the islands, and it is almost entirely in American hands. It is burdened here by the difficulty of procuring labour, and at San Francisco by a heavy import duty. There are thirty-five plantations on the islands, and there is room for fifty more. The profit, as it is, is hardly worth mentioning, and few of the planters do more than keep their heads above water. Plantations which cost $50,000 have been sold for $15,000; and others, which cost $150,000 have been sold for $40,000. If the islands were annexed, and the duty taken off, many of these struggling planters would clear $50,000 a year and upwards. So, no wonder that Mr. Phillips's lecture was received with enthusiastic plaudits. It focussed

[focused] all the clamour I have heard on Hawai'i and elsewhere, exalted the "almighty dollar," and was savoury with the odour of coming prosperity. But he went far, very far; he has aroused a cry among the natives "Hawai'i for the Hawaiians," which, very likely, may breed mischief . . and his hint regarding the judicious disposal of the king in the event of annexation, was felt by many of the more sober whites to be highly impolitic.

The reciprocity treaty, very lucidly advocated by Mr. Carter, and which means the cession of a lagoon [Pearl Harbor] with a portion of circumjacent territory on this island, to the United States, for a Pacific naval station, meets with more general favour as a safer measure; but the natives are indisposed to bribe the great Republic to remit the sugar duties by the surrender of a square inch of Hawaiian soil; and, from a British point of view, I heartily sympathise with them. Foreign, i.e. American, feeling is running high upon the subject. People say that things are so bad that something must be done, and it remains to be seen whether natives or foreigners can exercise the strongest pressure on the king. I was unfavourably impressed in both lectures by the way in which the natives and their interests were quietly ignored, or as quietly subordinated to the sugar interest.[81]

Table 1.3. Milestones achieved by the Hawaiian sugar industry prior to 1876

1825	First sugar plantation attempted in the Mānoa Valley of O'ahu
1835	First successful plantation started at Kōloa, Kaua'i, by Ladd and Co.
1837	Hawai'i's first sugar crop, 2.1 tons
1838	Twenty animal-powered and two water-powered mills in operation
1850	Royal Hawaiian Agricultural Society organized
1852	First sugar centrifugal installed in Makawao, Maui
1852	First Chinese brought to Hawai'i as laborers
1853	First steam engine installed at Kōloa, Kaua'i
1854	Lahaina cane variety brought from Tahiti by Captain Edwards
1857	First sugarcane irrigation ditch constructed at Lihu'e, Kaua'i, by W. H. Rice
1859	First steam mill in Hawai'i, at Lihu'e
1860	First mill on O'ahu at Kualoa by Charles H. Judd and Samuel G. Wilder
1861	Twenty-two sugar factories: nine powered by steam, twelve by water, one by animal
1863	First vacuum pan in Hawai'i at Pepeekeo
1868	First Japanese laborers arrive

Source: Gilmore, *Sugar Manual,* 1939.

The optimism and ferment that Bird noted were soon rewarded when the Reciprocity Treaty was ratified, and in 1876 the tariff was eliminated. The number of Hawaiian sugar plantations quickly jumped from 33 to 46, and the industry aggressively moved into a new and, at least temporarily, prosperous era (see Table 1.3).

Summary

Like many other food plants, sugar was brought to the Hawaiian Islands by its Polynesian settlers, and its numerous varieties were used for many purposes. After the islands were "discovered" by British Captain James Cook, non-native settlers began to establish themselves and exert a degree of influence on the Hawaiian monarchy. The islands' well-established and strictly enforced systems of land tenure and water rights based on absolute control by the king evolved under the influence of foreign trade and immigration. Early immigrants recognized that sugarcane could be produced easily—but Hawaiian labor would be required. The earliest commercial success came in the late 1830s, when Kōloa Sugar Company was established on Kauaʻi, employing Hawaiian and immigrant workers in a paternalistic system based on wage labor, plantation housing, and a company store.

The Hawaiian monarchy began allowing private land ownership under the provisions of the Great Māhele in the 1840s, and by the 1860s a substantial amount of land had been purchased by common people. Land for crops could, in turn, be purchased by entrepreneurs, making development of commercial agriculture feasible. The California gold rush of the 1850s, as well the decline of the whaling industry and the start of the Civil War in the 1860s, provided incentives for entrepreneurs to expand commercial production and export of sugar, molasses, rice, and other agricultural products. The sugar industry grew throughout the islands, creating a growing demand for field and factory labor. However, the steady decline of the Hawaiian population (due mainly to exposure to diseases brought to the islands by visitors and immigrants) reduced the labor force, driving plantation owners to actively recruit foreign laborers, especially workers from China and Japan. Productivity increased as a result of the arrival of more productive sugarcane varieties like Lahaina and importation and local construction of more efficient and effective milling and sugar processing equipment.

Stimulated by high sugar prices in the 1850s and 1860s, a number of new sugar companies were formed throughout the islands. On Maui, sugar plantations and mills were established near Lahaina, Wailuku, and on the windward and leeward slopes of East Maui. However, falling sugar prices after the Civil War and protectionist duties on imported sugar reduced the profitability of the industry, and many sugar companies failed. The resulting economic downturn in Hawai'i caused the leadership of the sugar industry to lobby for either annexation of the kingdom by the United States or a treaty eliminating tariffs on trade. After several years of economic and political uncertainty, the Reciprocity Treaty was signed in 1876, eliminating most tariffs between the United States and Hawai'i. The treaty set the stage for rapid growth and development by the industry, both by increasing the prices it received for exports and by reducing the costs of imported manufactured goods.

2
Sugar Booms—1876 to 1897

In 1876 Hawai'i was home to 35 plantations, 12 of which were located on Maui: East Maui Plantation Company, 'Haikū Sugar Company, Grove Ranch Plantation, and Alexander & Baldwin were all located on the northwestern slopes of East Maui; Waiheʻe Sugar Company, Wailuku Sugar Company, Bal & Adams, and Waikapū Sugar Company had all been formed on the eastern slopes of West Maui during the heady days of high Civil War sugar prices; Pioneer Mill Company and West Maui Plantation were operating on the dry western slopes of West Maui; Hāna Plantation Company was the sole plantation on the wet eastern point of East Maui; and 'Ulupalakua (Makee) Plantation was struggling to grow cane at high elevations on the dry southwestern slopes of East Maui. Before 1876 the U.S. duty on Hawaiian sugar was 40 percent; passage of the Reciprocity Treaty eliminated the tariff on Hawaiian sugar entering the United States in exchange for use of Pearl Harbor as a refueling facility, effectively increasing sugar value to the plantation by a similar amount. As might be expected, this led to rapid expansion and reorganization of the industry. Sugar production skyrocketed from 13,000 tons in 1876 to 260,000 tons in 1897. On Maui, 14 new plantations were formed between 1876 and 1897, but a total of 18 eventually ceased operations, primarily due to mergers.[1] By the close of 1897, only five plantations were operating on Maui.

Sugar Industry Politics

As Isabella Bird had noted, in the early 1870s the sugar industry was struggling with labor shortages caused by the continuing decline in the Hawaiian population. Though the first Chinese laborers had arrived in the islands in 1852, the industry needed much more labor, and in 1875 it began seeking the cooperation of Great Britain to bring "Hindoo" (a phonetic misspelling of Hindu) labor from India. In addition, because of disappointing negotiations with the United States in the 1850s and 1860s, the Hawaiian industry began looking to Australia as a market for its sugar, proposing to send the entire crop of 1876–1877 there. If the United States did not want Hawai'i, perhaps the islands could become a British colony. Whether this was a serious possibility or not, "as soon as the United States became aware of this move, they readily seized the opportunity, and entered into the . . . Reciprocity Treaty, which allowed raw sugar to be imported into the United States free from any duty whatever."[2]

The stimulation of the Hawaiian sugar industry, first by increasing demand during the Civil War and later by the Reciprocity Treaty, allowed the industry to adopt better practices in its fields and factories. Yields increased fourfold from 1850 to 1882, though net profits per acre improved by only one-third (see Table 2.1). Nevertheless, the industry was proud of its progress. In 1882 sugar industry leaders quoted a speech made by Judge William L. Lee more than 30 years earlier, predicting that by 1880 Hawaiian valleys would be "blooming with coffee and fruit trees, our barren hill-sides waving with luxuriant cane fields, our worthless plains irrigated and fruitful, and the grass huts now scattered over our land, replaced by comfortable farm houses." The industry leaders noted that, with the exception of the coffee and fruit trees, his prediction had been "verified to a remarkable degree by the present condition of the country."[3]

The year 1882 was important to Hawai'i's sugarcane industry for several reasons. Industry leaders were concerned with the uncertain future of the Reciprocity Treaty, which was due for review in 1884. Hawai'i's system of immigrant labor contracts was under attack by sugar interests in the United States, who charged that the contracts represented "a species of human bondage." In response to this threat, a meeting of "the Sugar Planters of this Kingdom" was held in March 1882, and its presentations provide a valu-

Table 2.1. Typical per-acre costs and returns of raising, processing, and shipping sugar in 1850 and 1882

	1850	1882
Sugar yield per acre (tons)	1	4
Cost of raising one acre	$12.00	$100.00
Cost of manufacture, shipping, etc.	$50.00	$248.00
Gross receipts, sugar & molasses	$236.00	$580.00
Net profits	$174.00	$232.00

Source: The Planters' Monthly, 1883, 116–117.

able snapshot of political and labor concerns as seen through the eyes of plantation owners.[4]

Faced with powerful opposition from mainland economic interests, the planters proclaimed that the Reciprocity Treaty was "of vital importance to the Hawaiian Kingdom, and its continuance is dependent in a great degree on the immediate action of the Hawaiian Planters." The supposedly "malicious slanders" of the sugar industry in the United States "should be boldly met and exposed," and "nothing but an organized and systematic effort on our part . . . will be of any avail."[5]

In response to these and other concerns, the Planters' Labor and Supply Company was organized in 1882, succeeding the Royal Hawaiian Agricultural Society with the objective to improve communication and coordination within the islands' sugar industry. The Planters' Labor and Supply Company was interested in all aspects of cane production and processing. Almost immediately, it began publication of *The Planters' Monthly* and formed committees to discuss and report on issues of labor, cultivation, machinery, legislation, transportation, and manufacture of sugar.[6]

In order for the Reciprocity Treaty to finally be renewed in 1887, the Kingdom had to concede the use of Pearl Harbor to the United States, a step that injured the pride of the monarchy and native Hawaiians. Yet another insult came in 1890, when the United States implemented the McKinley Act, which removed the tariff on raw sugar entering the States from any country. And to stimulate the domestic sugar industry, the U.S. government decided it would pay a 2 cent per pound "bounty" to U.S. sugar producers. This threatened to wreak havoc on the Hawaiian industry by negating its advantage over other foreign producers.

To complicate matters even further, before the McKinley Act could take effect on April 1, 1891, King Kalākaua died while on a trip to San Francisco, and his sister, Queen Lili'uokalani, took power. Political intrigue soon increased in Hawai'i. As the sugar industry began to suffer from the effects of the act, powerful interests started to view annexation of the Kingdom by the United States as a viable option: it would allow Hawai'i to receive the U.S. bounty for domestic sugar, thus helping to stabilize the islands' political and economic situation. Revolutionaries, including some prominent sugar industry leaders, staged an almost bloodless coup in January 1893, and, assisted by the U.S. Navy, overthrew the monarchy. A provisional government led by Sanford Dole, who had been president of the Planters' Labor and Supply Company in 1885–1886, took power. On July 4, 1894, after putting down a counter-coup and convening a constitutional convention, Dole proclaimed the Republic of Hawai'i with himself as president. The republic was promptly recognized by the United States and Great Britain.[7]

The McKinley Act was repealed in 1894, removing the bounty on domestic sugar, and the United States implemented the Wilson-Gorman tariff, which, in conjunction with a new reciprocity agreement, restored Hawaiian sugar to its pre-1890 tariff-free status. However, momentum for annexation remained strong, and in September 1896 the Hawaiian Senate ratified a treaty of annexation. Nevertheless, it took almost two years and much debate before the U.S. Senate concurred in July 1898, just months after hostilities in Cuba and the Philippines initiated the Spanish-American War. Hawai'i was now a territory of the United States.[8]

Plantation Labor

The period between the Reciprocity Treaty and annexation by the United States brought rapid societal change in Hawai'i. The strong economy of the islands, growing U.S. influence, labor shortages, and fluctuations in immigration converged, resulting in political, economic, and social damage in the form of official corruption, extravagant consumption by some foreigners, and increased smuggling, alcohol, and opium use among the people of Hawai'i. Many native Hawaiians left their small, isolated farms and villages for life in the larger towns and cities, both to sample urban life and because the expanding sugar and rice plantations had begun to encroach on their small farms. Their decreasing population and urban migration made ethnic Hawaiians increasingly irrelevant to the sugar industry; consequently, thousands of Chinese, Japanese, Germans, Norwegians, Portuguese, Spanish, and South Sea Islanders were brought to the islands as contract laborers. In 1879, for instance, a total of 4,150 men, women, and children arrived, more than 3,500 from China alone.[9]

In 1882 the Planters' Labor and Supply Company Committee on Labor reported that "the question of labor is, without doubt, the most serious problem" that the industry faced. Revealing their ethnic prejudices, the planters had long discussions on the relative merits of ethnic Hawaiians and different immigrant groups to the industry. The committee indicated that "concerted action on the part of all the planters" to secure more foreign contract laborers had not occurred because of "the difference of opinion as to the nationality of laborers best adapted to our wants." There was general agreement that "the native Hawaiian is naturally superior as a plantation hand to any of the importations hitherto made." Although ethnic Hawaiians had "supplied nearly the entire labor force of the kingdom" in 1872, by 1882 they represented only about 24 percent of plantation laborers and their percentage was steadily decreasing.[10] Bottom-line profits and underlying ethnic stereotypes figured heavily in every decision made by the sugar plantation owners. Ethnic prejudice was rampant, and the management decisions of industry leaders reflected their perceptions of the ethnic groups that made up their workforce. Moreover, these perceptions changed over time. For example, the Committee on Labor reported that in 1872 the planters favored Japanese laborers; a decade later, though, they seemed to prefer the Chinese and Portuguese. It was "generally conceded that Chinese laborers were best used in the boiling-house or for routine hand field work and irrigation. In contrast, the Portuguese were better suited to be in the many mechanical industries pursued upon the plantation," but they did not "as a rule, take kindly to the monotonous hand work in the field, or succeed in the handling of teams." This was better left to "the native Hawaiian . . . with his natural fondness for the horse and aptitude in handling our half broken cattle." In addition, ethnic Hawaiians were willing to work "without regard to regularity of hours or time of day." Nevertheless, the committee recommended that the Kingdom seek discussions with Great Britain about bringing "Hindoo" immigrants from India, but only if they came "under the sole protection of Hawaiian laws," a requirement that flew in the face of British demands that workers from their colonies by protected by British law.[11] As a result, no labor was imported from India.

Ethnic stereotypes were common and planters freely exchanged their perceptions of the different groups that made up the workforce. Colonel Z. S. Spalding, former U.S. Consul in Honolulu and manager of the Makee Plantation on Kaua'i, summarized his personal feelings as follows:

(1) *Natives of the island—The best labor for general purposes—quiet and peaceable; active and quick to learn; easily fed and provided for, but shiftless and improvident of themselves. They require mild to firm discipline, and to be kept from the temptation of liquors, etc.*

(2) *Portuguese—Strong and good for hard work, but not easily managed. They require expensive food, if fed by the employer, but live cheaply when left to themselves and allowed ration money.*

(3) *Chinese—Good labor for most of plantation purposes when the supply is sufficient to prevent their taking advantage of the necessities of the employer. As day laborers, at high rates of wages so that they are not*

obliged [to] work regularly, they are unsatisfactory and hard to deal with.

(4) *Japanese—Not so quick and apt as the Chinese, but quiet and d[o]cile, and steady workers—especially in field work, with hoe, etc. They prefer their own style of food, and, like the Chinese, do not eat meat enough to become strong.*

(5) *Natives of the Southern Island—But few in the country; some of them are indolent and worthless; others are next best to our native laborers; and*

(6) *Germans, Norwegians, etc.—I have had no personal experience with these people, as contract laborers, but have employed a few for special purposes. They are, I think, a failure as plantation laborers.*[12]

In 1886 U.S. Consul-General J. H. Putnam reported: "The total population of the Kingdom at the last census (1884) was 81,000, of which 42,000 were natives, 18,000 Chinese, 9,377 Portuguese, 2,066 Americans, 1,200 English, and the remainder distributed among various nationalities." Two years later, the number of Japanese had increased through official immigration to about 2,500, and the number of Chinese had grown to more than 20,000. Putnam reported that because of the strong demand for labor on sugar plantations:

> . . . a board of immigration was instituted, and agents sent to Norway and Sweden, the South Sea Islands, and Portugal. Men were employed on three years' contracts by the Government and re-contracted to the planters. Within the past year Japan has also been added to the countries from which laborers are to be employed, and two ships have arrived at the port of Honolulu with Japanese immigrants, who are contracted to the planters. By this means the demand for labor has been partially met, but the Government is still extending its efforts in Japan and Portugal to meet the increasing requirements of the planters. The contracts . . . generally embrace free passage to the Kingdom, advance wages, board, lodging, medicine, and medical attendance, and are for three years, at an average salary of $9 per month for males and $6 for females. . . . The employer and employe[e] are compelled to fulfill their contracts; laborers who desert are arrested, fined and returned, the same as are deserters from merchant ships. Over these foreign laborers the Government undertakes a paternal care and guardianship, and, through inspectors, their condition and treatment are constantly investigated, and when they are improperly used they are withdrawn and re-contracted.[13]

Compensation for unskilled laborers depended on their nationalities and reflected the plantations' willingness to pay more for ethnic Hawaiian labor.

- Ethnic Hawaiians—$20 to $25 per month without room or board
- Portuguese—$9 per month with board, or an allowance of $7 for food
- Japanese—$9 per month and $6 for food
- Chinese—$10 per month and $6 for food

In addition, for new immigrants the plantation had to pay for their passage to Hawai'i. For Portuguese laborers, the cost was $125; for Japanese, it was only $60.[14]

In 1886 most plantations were providing payment for food rather than supplying it directly to the workers. The plantations also supplied medical care but not clothing. Single men were housed in bunkhouses, and families received one or more rooms in plantation housing, normally free of charge. Some valued workers received small cottages with space for a garden, usually at nominal rates.[15]

By 1890 nearly 19,000 laborers were employed on Hawai'i's plantations, approximately half as day laborers and half as contractors. Of the total, over 8,600 were Japanese, just over 4,500 were Chinese, 3,000 were Portuguese, almost 1,900 were ethnic Hawaiians, and the remainder (less than 1,000) were South Sea Islanders, Americans, British, and others. Almost 1,500 were women, mostly Japanese, and more than 400 were minors, mostly Portuguese. The Japanese primarily worked in teams as contractors, taking advantage of the higher wages and freedom of supervision by plantation staff that contractors enjoyed. In contrast, the Chinese were primarily day laborers.[16]

Labor continued to be a problem for many plantations in the 1890s. Labor availability and costs were major constraints on plantation profits, and the very different perceptions and treatment of ethnic groups complicated plantation management. In general, wages of the four largest ethnic groups differed significantly (see Table 2.2).

The Republic of Hawai'i's Department of Foreign Affairs also reported on typical plantation labor costs, revealing that wages of unskilled field labor were far below those of skilled and supervisory labor.

Engineers on plantations, from $125 to $175 per month, house and firewood furnished.

Table 2.2. Typical labor costs on Hawai'i's sugar plantations by ethnic group and type of labor

Ethnic group	Skilled labor	Contract labor	Day labor
	Dollars per month		
Hawaiians	55	19	21
Portuguese	47	20	22
Japanese	41	16	19
Chinese	38	18	17

Source: Thrum, *Almanac,* 1890.

Sugar boilers, $125 to $175 per month, house and firewood furnished.

Blacksmiths, plantation, $50 to $100 per month, house and firewood furnished.

Carpenters, plantation, $50 to $100 per month, house and firewood furnished.

Locomotive drivers, $40 to $75 per month, room and board furnished.

Head overseers, or head lunas, $100 to $150.

Under overseers, or lunas, $30 to $50 with room and board.

Bookkeepers, plantation, $100 to $175, house and firewood furnished.

Teamsters, white, $30 to $40 with room and board.

Hawaiians, $25 to $30 with room; no board.

Field labor, Portuguese and Hawaiian, $16 to $18 per month; no board.

Field labor, Chinese and Japanese, $12.50 to $15 per month; no board.[17]

Prior to annexation of Hawai'i to the United States in 1898, the contract labor system made it illegal for workers to strike while under contract. However, in a few instances (see Table 2.3), plantation workers attempted to take their grievances beyond local management. For example, in 1881 Norwegian workers at the Alexander & Baldwin Plantation on Maui and the Hitchcock Plantation on Hawai'i refused to work, objecting to the quality of food they were receiving, the terms of their labor contracts, and the requirement that they work more than 10 hours a day. But these protests were easily put down by the courts and police, and the workers were forced to return to work. Similarly, in 1891 Chinese workers on three plantations on the island of Hawai'i claimed to have been cheated by a labor contractor and refused to work. After a one-day strike, police attacked the strikers, broke windows in a Chinese labor camp nearby, jailed 55 laborers, and forced the remainder to return to work. In 1894 a

Table 2.3. Labor strikes on Hawaiian sugar plantations prior to 1898

Year	Plantation, island	Ethnic group	Issues	Outcomes
1841	Kōloa, Kaua'i	Hawaiians	Wages	No compromise, returned to work
1881	Alexander & Baldwin, Maui	Norwegians	Food quality	Jailed, returned to work
1881	Hitchcock, Hawai'i	Norwegians	Food, labor contract, working conditions	Workers lost court case, returned to work
1891	Kohala, Union Mill, Hawai'i; Hawai'i	Chinese	Contract labor dispute	Police attacked, jailed 55 workers
1894	Kahuku, O'ahu	Japanese	Contract labor dispute	Workers arrested, fined

Source: Takaki, *Pau Hana,* 1983.

protest march of Japanese laborers from Kahuku Plantation to Honolulu ended with the protesters arrested, fined, and forced to walk back to the plantation.[18]

Plantation Practices

As plantations expanded and became more profitable after the Reciprocity Treaty, owners invested in farming practices, infrastructure, and inputs that promised increased productivity. Of course, choosing the best available sugarcane land for expansion was important. In 1886 Colonel Spalding described the islands' best soils from the perspective of sugar planters. These were on the leeward sides of the islands, "where the washings from the mountains have settled on the plains below." Planters preferred "a clayey soil, sufficiently porous for proper drainage, but fine enough to hold necessary moisture." Soils that were "intense red, like iron rust" or black were generally better than "bright yellow" or "bluish gray" soils. In addition, lands that were too sandy or received salt spray from the sea were not good for cane. While land values varied, the market value of the best sugarcane lands was $25–50 per acre; the best pasture land was $2–5 per acre; and the best wooded land was $1 per acre.[19]

Spalding also enumerated the equipment that plantations needed, most of which was imported from the United States:

> . . . plows, steam and horse power; harrows, hoes, shovels, picks, mattocks, axes, knives (cane), hatchets, etc., for the field work; locomotives, cars, wagons, carts, sleds. Flumes, etc., for the transportation of cane to the mill, etc.; engines, mills, boilers, clarifiers, evaporators (open and vacuum), vacuum pans, c[e]ntrifugals, etc., for the manufacture of the sugar; and bags, kegs, barrels, etc., for packing and shipping products. The carpenter's and black-smith shops, and other repairing departments, must be kept supplied with all kinds of tools, ropes, iron, lumber, etc.; and wire, posts, nails, staples, etc., for fencing.

He complained that the cost of these items in Hawai'i was 25–50 percent greater than in New York or San Francisco, and "the cost of machinery and tools for a plantation (favorably located) to make ten tons of sugar per day during the season will average . . . $75,000 or $100,000."[20]

As a broad generalization for an "average" plantation, the total cost of producing and manufacturing a ton of sugar was about $80–90. This included "about sixty days labor, besides the animals or other power used in plowing, etc." which cost about $50 per ton of sugar produced. The costs of transporting cane to the mill and mill expenses, including fuel, bags, and labor, added another $30–40 per ton of sugar.[21]

Plowing, Planting, and Cultivation

In an attempt to spread better management practices throughout the industry, in 1886 the Planters' Labor and Supply Company's Committee on Cultivation requested information from "those who have made a special study of any of the following subjects"—lands or soils best adapted to cane, fertilizers, plowing, irrigation, economy of labor, cane planting, stripping of cane, and harvesting. They indicated that "any labor saving machinery will be greeted with approval by planters. . . . This is particularly true in view of the difficulty of obtaining laborers and the high cost of those who can be procured."[22]

Steam plows were just being adopted by the industry in the 1880s, with East Maui plantations taking a lead role in their importation and use. R. H. Fowler, representative of the English steam tractor manufacturer John Fowler & Co. Ltd., wrote, "Now that the Mechanical success of Steam

Figure 2.1. Steam tractor attached to cable-drawn turning plow.
Photo courtesy of HARC

Ploughs in these islands has been proved by actual trial, both on the plains
at Kahului [Maui], and on the steep side hills at Hāmākua [Hawai'i], I wish
to draw attention to the many advantages steam cultivation has over horse
or animal cultivation." A set of two 16-horsepower compound steam engines was used at Kahului, "burning about a ton of coal per day." Pairs of
steam engines, one on either side of the field, were used for plowing. Each
engine had a large diameter winding drum turned by the steam engine. Onto
the drum a long length of wire rope was wound, which was used to pull
the plow or cultivator back and forth across a field. Each "balance" plow
was fitted with two sets of moldboards facing opposite directions. When
the plow was pulled by the wire cable in one direction across the field, one
set of plows was engaged. When the plow reached the edge of the field, the
set of plows was tilted, and the other set of moldboards was engaged, turning the soil as the plow was pulled back across the field (see Figure 2.1).
With a five-row plow, the system could plow 15 acres or cultivate 25 acres
per day. Fowler pointed out numerous advantages of steam plowing and cultivation, including reduced cost of labor and draft animals. In addition,
Fowler claimed that "ploughing by steam breaks up the land more efficiently
than two or three ploughings with animals," and plantations could prepare
land more quickly and plant in a timely manner. Finally, since draft animals
are not needed, "all treading and compression of the soil and sub-soil

[by animal hoofs] is thereby entirely avoided, and the implement [plow or cultivator] is drawn at a much more rapid pace, throwing up the soil to a greater depth and in a loose state, enabling it to derive full benefit from the influence of the atmosphere."[23]

Charles Notley, of Paauilo, Hamakua, described the pair of steam-powered plows in use on Hāmākua Plantation on Hawai'i. The plantation used "two engines each of 8 horse power, and weighing 12 tons each when in full working order with drums and rope. There [was] one water cart, all iron and fitted with a pump, a balance plough four feet wide which [cut] four furrows at once, and one cultivator six feet wide with seven tines." The steam engines used on the Hāmākua Plantation were smaller than those used on Maui, possibly because the cane fields on Hawai'i were steeper than those on East Maui.[24]

J. M. Horner, an American agriculturist turned sugar cultivator in Hawai'i, recommended that cane land be plowed twice and harrowed twice prior to planting, with the second plowing just before planting to kill all weeds. A furrow plow was recommended for making "seed furrows" five to six feet apart. Horner suggested running "a one-horse subsoil plow in the bottom of the furrow just ahead of the planters, loosening the soil from two to six inches, making nice, fresh seed-bed, the seed to be planted under the bottom of the furrow in mellow soil."[25] (See Figure 2.2.)

Figure 2.2. Planting cane by hand. Photo courtesy of HARC

In 1886 Colonel Spalding gave a clear description of how cane seed pieces were planted, how the buds began to grow, and how the young crop was tended. Pieces of cane stalk, or joints, were placed in furrows that had been plowed. "When the joint is planted it is stripped of its leaves . . . , and where the leaf had been previously attached the roots spring out." The buds then begin to grow, each joint producing one bud with a leaf that "completely envelopes the stalk, covering and protecting the bud or eye. . . . When the leaf [of the new shoot] is young and green it is closely wrapped around the stalk; as it matures it unfolds itself, and as the stalk grows upward and matures the lower leaves wither and die." Each bud can produce a stalk. Spalding concluded, therefore, that "the best cane for planting is, of course, that which is shortest jointed, as having the most eyes to a given length." Cane with long joints was considered best for grinding because most of the sugar was located in the long internodes of the joints.[26]

Seeking to reduce the hand labor associated with planting, C. C. Coleman of Honolulu invented a cane planting machine which, he claimed, would do the work of ten men. It was mounted on wheels and drawn by two or four animals. This machine opened the furrow, dropped the cane seed piece, and buried it, all in one pass. It was tested on Oʻahu, and *The Planters' Monthly* indicated that "another trial of the machine is shortly to be made at Spreckels' [HC&S] Plantation, Maui." Such machines were soon adopted where land was not too rocky or steep for their use.[27] Horner recommended that cultivation for weed control should begin as soon as the cane leaves emerged from the soil after planting. A horse- or mule-drawn cultivator was preferred over hoes because it was much more economical and it produced better results. Horner estimated that one man with a horse or mule could cultivate four acres a day, while it took five men to hoe just one acre a day. Cultivation was repeated whenever weeds began to grow and it continued until the cane was too large to allow the implement to pass. On fields that were not irrigated and did not require that an irrigation furrow be maintained, each pass of the cultivator caused a little soil to "roll down" around the cane, and the field would be nearly level by the time of harvest.[28]

Spalding recommended that cultivation of unirrigated cane should be "done by means of small plows and cultivators drawn by one horse, or a mule, and some hoeing." If irrigated, the cane had to be planted in deep furrows "in order that water may be run in as required." As a result, weeds had to be controlled with hoes, "as plows and cultivators would tear down the sides and banks" of the irrigation furrows. As the cane grew, new leaves emerged near the top of the stalk, and older leaves withered and died from the bottom upward. These dead leaves were "removed, or stripped from the stalk to admit air and light—necessary to the proper maturing of the cane. This is generally done two or three times in a season."[29]

Fertilizers

In 1888 J. M. Lydgate, a sugar planter at Laupāhoehoe, minister, and botanist, gave the Committee on Fertilizers his recommendations, saying that "already many estates have reached the condition from which they say, 'we must fertilize if we would get a crop.'" Even though bone meal had been the most commonly used fertilizer in the industry, "the phosphoric acid that the plant needs . . . is so securely locked up in the hard particles of bone that it is not available in time to be of any benefit to the cane. For this reason bone meal has fallen into general disfavor throughout the world, and is little used by intelligent farmers."[30]

Lydgate's committee recommended using either bone superphosphate made by treating bone meal with acid to make the phosphorus "immediately available" or "the so-called phosphatic guanos of the South Sea Islands. . . . The South Sea guanos, or at least those from Baker's and Howland's Islands, are of an excellent quality as phosphates, and merit our attention as a cheap and valuable source of fertilizer close at hand." With regard to nitrogen, the committee was not aware of any plantation use of sodium or potassium nitrate or ammonium sulfate. The committee members recommended Peruvian guano, though they feared that it was no longer easily available because the supply had been exhausted. They also reported that "fish-scrap," available at $10–15 per ton, was an "efficient and economical fertilizer" that had been used in Lahaina. Wood ashes were recommended as a good source of potassium, and the committee speculated that potassium hydroxide, if mixed with bone meal or other organic sources of phosphorus like manure or filter press mud, might be a good source of both phosphorus and potassium. But in the end, Lydgate concluded that "as a result of two or three years' experimenting, it seems to me nothing is so uniformly certain of giving good results as stable manure, filter press mud, and ashes." The problem was that supplies were limited, and the materials were bulky and hard to transport and apply. "To undertake to manure 500 acres, and do it properly, means the exhaustion of the whole available supply of the country, and a herculean task of transportation and application." Finally, Lydgate warned that "fertilizers are subject to the grossest fraud, and very often the price bears no relation whatever to the value of the article, and its pretended analysis none to its real composition." As a result, "no fertilizer should be bought without a definite knowledge of its composition, or a chemical analysis."[31]

Harvesting

Most of the cane in the islands flowered in November if the crop had been planted in the previous winter and was old enough for flowers to be induced by the shorter days of autumn. Spalding noted that "after flowering [the stalk] stops growing, the top dies, and it sends out side shoots from the eyes near the upper joint. This is detrimental to its sugar-producing qualities, and therefore the canes are to be cut as soon as convenient after the flowering season." The cane that was planted in the late spring or summer was too young to flower in November and continued to grow until it was harvested during the next winter. Such fields could "produce large crops to the acre by having long seasons of growth—say sixteen to twenty months."[32] The cane was harvested by cutting the stalks with "hatchets or knives, near the bottom—in fact, just under the ground, if possible." The stalks were then "stripped of their leaves and worthless tops" (see Figure 2.3).[33]

Figure 2.3. Cutting and piling cane by hand. Photo courtesy of HARC

Figure 2.4. Loading cane onto railcars by hand. Photo courtesy of HARC

After the cane was cut and piled on the ground, it was loaded by hand onto carts (see Figure 2.4) and pulled by bullocks, mules, steam tractors, or "steam tramways" (narrow-gauge steam railways) for transport to the factory.

On Maui, in addition to using carts pulled by animal traction, the Spreckels Plantation, Hāna Plantation, and Olowalu Plantation used steam tramways to haul cane. On the rainy windward sides of the islands, most of the plantations used "wet flumes" to move their cane down from the fields above the mill. Like irrigation ditches, these wet flumes took water from streams and used it to float harvested cane either all the way to the mill or to a transfer station where it was discharged into a railcar for transport to the mill. Spalding's Makee Plantation on Kauaʻi used flumes to move cane to the mill from fields up to three miles away, and he judged fluming cane "the best and least expensive method known," explaining, "the flume is built either in a $\vee$ shape of boards placed at right angles, or with flat bottom and V sides, as $\sqcup$, and supported on trestles with a grade of fall varying from 1½ to 3 feet in a hundred." Spalding indicated that his flumes could convey "40 to 50 tons of cane to the mill per hour, though [they were] seldom . . . used to full capacity." On Maui, Alexander & Baldwin Plantation, Wailuku Sugar Company, and Pioneer Mill Company all transported part of their cane with wet flumes (see Figure 1.2).[34]

The Planters' Labor and Supply Company's Committee on Transportation of cane concluded that flumes were "the best and cheapest medium," followed by rail, then "mules, cattle and traction engines." But R. H. Fowler, in a letter to *The Planters' Monthly,* extolled the virtues of narrow gauge railroads with portable tracks to haul cane to the mill, the reasons being "cheapness, portability, lightness and strength." He recommended a maximum gauge, or track width, of 20 inches for small mills, allowing cane cars to carry 3,000 pounds of cane in hilly lands. For larger mills he suggested 24-inch tracks because of their ability to carry larger locomotives and longer trains of cane cars.[35]

Factory Operations

The Planters' Labor and Supply Company's Committee on Mill Machinery reported on new factory technologies, and Hawaiian Commercial & Sugar's (HC&S) Puʻunēnē mill was clearly an industry leader. Its five-roller mill was the first of its kind in the islands and had a hydraulic attachment to assure that the same pressure was exerted by the rollers regardless of the amount of cane passing through them. The resulting bagasse had only 38 percent moisture and could be burned immediately without further drying. The committee concluded that "although the largest mills crush nearly if not quite as efficiently as this, it seems that with the five-roller arrangement, a much larger amount of cane can be put through in a given time than with the triple rolls of the same size." HC&S was also using a much more efficient "compound" steam engine to drive its mill, extracting so much energy from the steam that "there is no more exhaust escaping from these engines than barely sufficient to heat the juice before it passes into the cleaning pans." Improved juice strainers and continuous "cleaning pans" (used to add lime and precipitate solids) were also noted. The "double and triple effect vacuum evaporators that had come into the industry since 1880 used ¼ to ⅙ the amount of steam as the older open evaporation pans, and because the evaporation occurred at lower temperatures, the juice was not discolored."[36]

By the early 1880s the double effect vacuum evaporation systems usually consisted of "two horizontal or vertical boilers, fitted with a drum of large heating surface. The juice in the first pan [was] boiled by exhaust steam under low vacuum and a temperature of about 185° [F]. The vapor rising from the juice of the first pan [entered] the drum of the second pan, and boil[ed] the juice at very high vacuum [at about 135° F]." The 8,450 gallons of water discharged from the vacuum pump for each ton of sugar produced could be collected in a cistern, cooled "by means of the wind," and "used over and over again," reducing the amount of fresh water supply needed by the mill.[37]

Honolulu Iron Works manufactured much of the special equipment required by the industry. For example, in 1882 it fabricated a 15-ton mill, complete with steam engine and gearing, and an "upright double effect" vacuum evaporation system for Pioneer Mill Company. For Waiheʻe Sugar Company, it built a steam engine for the mill and gearing "for driving centrifugals," as well as a pair of six-foot "compound boilers and one flue boiler."[38]

Plantations

The period between implementation of the Reciprocity Treaty in 1876 and annexation of Hawaiʻi by the United States in 1898 was turbulent for the Maui sugar industry. Twelve plantations were operating in 1876, and 14 plantations were formed between 1876 and 1897 (see Table 2.4). But, as mentioned at the opening of this chapter, most of Maui's plantations failed or merged with others, many as the result of the depression of 1891–1894 that coincided with low sugar prices caused by the McKinley Act. As of 1897,

Table 2.4. Plantations operating on Maui, 1876–1898

On the dry, leeward western and northwestern slopes of East Maui	
East Maui Plantation Company	1850–1886
ʻHaikū Sugar Company	1858–1883
ʻUlupalakua (Makee) Plantation	1858–1883
Grove Ranch, Plantation	1863–1889
Alexander & Baldwin	1872–1883
Hāmākua Plantation	1878–1880
Hawaiian Commercial & Sugar	1878–present
Lilikoe Sugar Plantation	1878–1880
Piiholo Sugar Plantation	1879–1880
J. M. Alexander	1882–1884
Pāʻia Plantation Company	1883–1821
Brewer & Associates	1884–1886
East Maui Stock Company	1884–1889
On the wet, windward side of East Maui	
Hāna Plantation Company	1850–1904
Kīpahulu Sugar Company	1879–1925
Reciprocity Sugar Company	1883–1898

Table 2.4. *Continued*

On the eastern side of West Maui

Waiheʻe Sugar Company	1862–1894
Wailuku Sugar Company	1862–1988
Waikapu Sugar Company	1862–1894
Bal and Adams	1865–1877
Bailey Brothers	1884–1884
Barnes & Palmer	1884–1886

On the dry, leeward western side of West Maui

Pioneer Mill Company	1863–1999
West Maui Plantation	1871–1878
Olowalu Company	1881–1931
Y. W. Horner, Planter	1883–1891

Source: Dorrance, *Sugar Islands,* 2000.

In 1886 Colonel Z. S. Spalding described how the Makee factory on Kauaʻi, which he managed, processed cane into sugar.

The cane is delivered on the cane carrier of the three-roller mill, and, with the aid of two men to feed, it passes through the rollers of the same under heavy pressure, and the juice; to the extent of from 50 to 65 per cent, is extracted. Some of us then pass the bagasse or cane trash through a second (two-roller) mill, where from ten to fifteen percent more juice is extracted. This is called double crushing. If water is used on the bagasse [to extract more sugar] before the second grinding it is called maceration. As the juice is extracted it is pumped through the heaters into the clarifiers, or large iron tanks, with copper or brass steam-coils, in which the juice is heated to a temperature of 200 to 210 degrees and lime added to correct the acidity and aid in defecation [settling of solids from the juice]. Afterwards the juice is thoroughly cleaned, by means of heat, and the water evaporated from it until the concentrated liquor stands at about 26 to 30° Baume [specific gravity = 1.22 to 1.26], when it is ready for the vacuum pan. In the best works this concentration is done in either the double effect or triple effect [evaporator], where the steam or vapor from the first pan is made to do the boiling in the next, and so on, the whole set working under vacuum of different degrees. In my works I practice concentrating before cleaning, as I find it easier to defecate or clean the heavy liquor than the light juice.

The work of the vacuum pan is to take in a certain quantity of this concentrated liquor at a temperature of, say, not above 150° Fah.[renheit], and to boil it down to what is known as the striking point, i.e., about 40° Baume [specific gravity = 1.38]. When the sugar boiler thinks he has about the right quantity and density, he admits a charge of cold "liquor," which has the effect of separating the small particles of saccharine; or crystallizable sugar, and is called starting the grain. After proper boiling, or concentration, another charge of cold liquor is taken into the pan, and the effect is to build up or enlarge the grain so started, and the operation is continued until the vacuum pan is sufficiently filled with sugar of the required grain [size]. Closing in, or finishing the strike, consists in boiling the mass to such a state or consistency as is best for drying, and the vacuum is then broken, and the whole mass discharged into a receiver or mixer, ready for the driers.

The drying is done by centrifugal motion, which separates the molasses, by whirling it through fine wire cloth, or screens [of a centrifuge], and leaving the dry sugar in walls on the inside. The dry sugar is then put in bags, or other containers, ready for the market; while the molasses is reboiled in the vacuum pan, and made into sugar of lower grade.[39]

just before annexation, Maui had only six operating plantations: Hawaiian Commercial & Sugar Company and Pā'ia Plantation Company on the western slopes of East Maui, Hāna Plantation Company and Reciprocity Sugar Company on the extreme eastern shores of East Maui, Wailuku Sugar Company on the eastern slopes of West Maui, and Pioneer Mill Company on the western side of West Maui.[40]

Sugar and Irrigation Ditches on Maui

Irrigation had been practiced for centuries by Hawaiians to irrigate their kalo (taro) fields. Building on this history, William Harrison Rice, founder of Lihu'e Sugar Plantation on Kaua'i, used Hawaiian laborers to build the Rice Ditch, a 10-mile ditch about 2.5 feet deep and 2.5 feet wide. Though it was unlined, leaky, and unreliable during dry periods, it made an impression on other planters, like Samuel T. Alexander from Maui, who visited it soon after its completion. Alexander, who grew up in Lahaina, was also familiar

At the time of Hawai'i's annexation to the United States in 1898, the islands were producing about 225,000 tons of sugar each year. How did production increase to this level from a mere 13,000 tons in 1876? The answer is a combination of increased sugarcane acreage, irrigation water, and mechanization. By 1890 the kingdom boasted 46 plantations with their own fields and mills, 18 plantations without mills, and six mills that processed cane for others but did not grow their own. In the 1890s the island of Hawai'i had by far the largest acreage and production of cane, followed by Maui, O'ahu, Kaua'i, and Moloka'i (see Table 2.5).[41] The largest plantations included: Hawaiian Commercial & Sugar Company (Spreckelsville, Maui—9,000 tons); Lihu'e Sugar Plantation (Lihu'e, Kaua'i—8,000 tons); Makee Sugar Company (Keālia, Kaua'i—6,940 tons); Onomea Sugar Company (Hilo, Hawai'i—6,800 tons); Pā'auhau Sugar Company (Hāmākua, Hawai'i—6,000 tons); Hilo Sugar Company (Hilo, Hawai'i—5,600 tons); Hakalau Plantation Company (Hilo, Hawai'i—5,000 tons); Pā'ia Plantation Company (Pā'ia, Maui—4,486 tons); Waiākea Plantation (Hilo, Hawai'i—4,400 tons).[42]

Table 2.5. Sugarcane acreage harvested and sugar produced by island in 1890

Island	Planters	Plant cane	Ratoon cane	Total cane	Sugar
			acres		tons
Hawai'i	30	23,706	13,580	37,286	60,750
Maui	12	9,610	3,405	13,015	32,850
O'ahu	5	1,935	1,487	3,422	5,700
Kaua'i	16	5,905	4,431	10,336	27,940
Moloka'i	1	35	55	90	200
Total	64	41,191	22,958	64,149	127,440

Source: Thrum, Almanac, 1890.

with the stone-lined Lahainaluna Ditch on the western slopes of West Maui, which was used to irrigate crops at Lahainaluna School.[43]

East Maui

By 1876 it was clear to Maui planters like Samuel Alexander and Henry P. Baldwin that dry leeward East Maui had too little rainfall to support profitable sugar plantations, and their only hope of success was to bring irrigation water from the high-rainfall areas on the windward side of East Maui to leeward plantations. Fortunately, the optimism and profits promised by the Reciprocity Treaty induced planters to risk the large investments necessary for developing the necessary ditch systems. That summer Alexander and Baldwin, owners of Alexander & Baldwin Plantation, proposed construction of the Hāmākua Ditch to the stockholders of ʻHaikū Sugar Company. Based on a preliminary survey by his brother, James, Samuel Alexander described the proposed project in a letter dated September 1876:

> . . . I propose to bring water enough to irrigate from 2,000 to 3,000 acres of land. There is no question about the water supply from which I am going to draw being ample. The only question is whether the ditch I propose digging is big enough to carry what water I want. The ditch, including all the bends, will be about 25 miles long, and we propose giving it a uniform grade of 20 inches to the chain (60 feet). It is to be 6 feet broad on top—3½ on the bottom and 3½ feet deep. If this will not carry all the water we want, we can enlarge it. We have got a great many shallow and a few deep ravines to cross.
>
> . . . My theory is that where the digging is good (where there is good hard clayey soil without rock requiring blasting) it will be the best plan to follow round these gulches and only make a short flume over the bottom so that winter torrents can pass under. The Māliko gulch, however, running through our Plan, is by far the most formidable gulch we will have to contend with. It is 300 feet deep and 800 feet broad and the sides precipitous and consisting of solid rock. I propose to pipe the water across this gulch. Have a pipe made of boiler iron and riveted about 18 inches in diameter which ought to be, I suppose, about 1/10 inch thick at the top of the gulch and increase in thickness to 3/10 inch at the bottom of the gulch. This pipe I would propose to dub well with hot tar and pitch. . . . The length of the pipe which I propose to put across the Māliko gulch will be just 1,000 feet.[44]

The proposal was accepted, and a lease was obtained from the Kingdom of Hawaiʻi stipulating that the ditch had to be completed within two years. Otherwise, ownership of the project would revert to the Kingdom. The Hāmākua Ditch Company was formed in November 1876, with the ownership, costs, and division of water divided as follows: ʻHaikū Sugar Company, 45 percent; Alexander & Baldwin, 25 percent; H. T. Hobron (acting for Grove Ranch Plantation), 20 percent; and James Alexander, 10 percent.[45]

Despite the fact that Baldwin had lost his right arm in a mill accident in March 1876, he took on responsibility of supervising construction of the Hāmākua Ditch. When the project reached the Māliko gulch, workers hesitated to lower themselves down a rope to the bottom of the gulch. In response, Baldwin led the way. Holding the rope between his legs and with his left arm, he lowered himself into the gulch and the workers followed.[46]

Another entrepreneur who emerged as a result of the Reciprocity Treaty was Claus Spreckels, a California-based businessman who had made a fortune in sugar refining. Spreckels initially opposed the treaty, in part because the high quality of Hawaiian sugar allowed it to be sold for some purposes without further refining, competing with his refined sugar. However, in 1875, after discussions with Hawaiian business leaders, he reversed his position and became a supporter. Clearly understanding the opportunities resulting from duty-free Hawaiian sugar, he arrived in the islands on the very steamer that brought news that the Reciprocity Treaty had been signed. Moving swiftly, he used his substantial financial resources to buy up much of the islands' 14,000-ton sugar crop for 1877 before the price rose. Learning of Alexander and Baldwin's Hāmākua Ditch, he visited Maui and spoke with Alexander about the project. Upon his return to California in September 1876, he engaged an irrigation engineer to design a similar ditch for him. Returning to Maui in May 1878 at the end of a severe drought on the island, Spreckels' engineer surveyed and designed what would become known as "Spreckels Ditch" *makai* (down slope) from the Hāmākua Ditch of Alexander and Baldwin. He also purchased half interest in 16,000 acres on Waikapū Commons and leased 24,000 acres of crown lands on Wailuku Commons, resolving the problem of land on which to grow cane irrigated with water delivered by his planned ditch. Through a clever combination of friendship with and financial help to King Kalākaua, Spreckels was able to lease the Maui water rights needed to build his new irrigation system and plantation. The terms of his lease stipulated that if Alexander and Baldwin's Hāmākua Ditch was not completed within two years, it would revert to Spreckels. But Baldwin's team crossed Māliko gulch in 1878, and the project was completed on time at a total cost of $80,000, yielding up to 40 million gallons per day (40 mgd) from its 17-mile length.[47] J. W. Vandercook, in his 1939 history of the Hawaiian sugar industry, observed that the completed Hāmākua Ditch was brilliantly successful and immensely profitable. "The hot lowland plain was perfect sugar country. The land was smooth and all of it was easy of access and simple to cultivate. The cane, with sun above it and water at its roots, throve magnificently."[48]

Spreckels' new ditch, which would serve his new Hawaiian Commercial Company (also known as Spreckels) plantation, "intercepted the streams on the northern slope of Haleakalā . . . from Honomanū westward and convey[ed] them by means of ditches about thirty miles long with a capacity of 50 mgd, delivering the water to the plantation at an elevation of 250 feet above the sea." Also known as the 'Haikū Ditch, it was the lowest elevation ditch on East Maui and was "remarkable for the great lengths of forty inch syphon [siphon] pipe used in crossing the ravines and for the steep gradient employed in conveying the water."[49]

Spreckels quickly made other major investments in Central Maui, founding Spreckelsville around a large, multistacked sugar factory. The factory was a modern marvel with the first electric lights and 5-roller mill in the Kingdom, as well as an extensive plantation railroad. In 1882 the Hawaiian Commercial Company was reorganized and incorporated as a California corporation operating at Spreckelsville under the name Hawaiian Commercial & Sugar Company (HC&S). However, by 1885 Spreckels' heavy investments in HC&S landed him in financial difficulties, with debts of over a million dollars.[50]

While Spreckels was busy developing his Central Maui town, plantation, and mill, the other Central Maui sugar entrepreneurs were anything but idle. In 1880 Alexander and Baldwin replaced the Paliuli mill with the larger Hāmākuapoko factory at Pā'ia to grind cane for 'Haikū Sugar Company. In 1881 Grove Ranch closed its Paholei factory and began having its cane ground at the new Hāmākuapoko factory. Also in 1881 Kahului Railroad Company was incorporated and over the next five years extended its tracks to Wailuku and Lower Pā'ia, allowing shipment of sugar from Kahului.[51]

In 1883 Alexander left the 'Haikū Sugar Company and moved to California, where he continued his active role in the partnership with Baldwin. When Alexander moved, Baldwin became general manager of both 'Haikū Plantation and Alexander & Baldwin Plantation, which was incorporated under a new name, Pā'ia Plantation Company. In 1886 Baldwin bought East Maui Plantation and merged it into 'Haikū Sugar Company. The same year, Alexander negotiated a contract for American Refinery in California to refine the raw sugar of most of the Kingdom's sugar companies, and the following year Baldwin was elected to the Kingdom's legislature from the Reform Party, which favored a constitutional monarchy. In 1889 Grove Ranch Plantation merged with Baldwin's Pā'ia Plantation.[52]

By the late 1880s Claus Spreckels' Hawaiian Commercial & Sugar Company was the largest plantation in the Hawaiian Islands. In 1886 its 1,600 acres of plant cane had yielded 5 tons of sugar per acre, and its 1,650 acres of ratoon cane produced 3.65 tons of sugar per acre. The plantation was in the process of clearing, plowing, and planting an additional 1,000 acres of land that would produce its first crop in 1888.[53] That year *Harper's Weekly* published a story about Spreckels' plantation, noting that it had yielded as much as 14,000 tons of sugar in a single year, and the average production was 6,000–8,000 tons. The article highlighted the importance of the plantation irrigation system, stating that "without the water obtained from the mountain-tops much of the valley would be an arid desert." The water from "a thousand springs" was conducted by flumes, "merely plank troughs four feet wide by about three feet deep" that carried the water "through tunnels, over deep gulches, along high trestles" to the plantation. The flume water emptied into plantation ditches that "traverse[d] the plantation in every direction, crossing and intersecting each other" as they carried the irrigation water to the cane fields. The head "ditch-minder" earned $115 a month and had 20 assistants earning $60 a month. They were supported by "large gangs of temporary employees" who stopped leaks and made other repairs. The author also reported seeing "a gang of about twenty men and women galloping up and down the ditches, the horses' hoofs compacting the soil and preventing to a great extent the water's seeping away into the earth."[54]

The *Harper's Weekly* story also extolled advanced technology that Spreckels had installed on the plantation, including "huge steamploughs that go bowling over the miles of fields" and "turn over more ground in a day than the ordinary ploughs on other plantations turn in a week." Spreckels' plantation railroad brought to the mill "in a single day" what oxen "would take a month to bring." The railcars were unloaded directly onto the mill cane carrier, and the factory expressed 770 pounds of juice and 147 pounds of raw sugar from each 1,000 pounds of cane. Sacks filled with the sugar were carried by rail "to the seashore, half a mile away, where there are anchored fleets of white-sailed vessels ready to carry the tons of sweetness to far-off America." The plantation relied on a workforce of 1,400 men—100 of them white and the remainder of the force "a mixture of Japanese, Chinese, South Pacific Islanders and Hawaiians." However, the high cost of sugar production in Hawai'i and political uncertainties caused Spreckels to look into sugar beet production in California. According to *Harper's Weekly,* "after several years' research and experiment," Spreckels expressed "confidence in the ability of California to grow sugar-beets" and he already had "under way works for the production of sugar from beets." If this effort were to prove successful, imports of sugar to the West Coast from the Kingdom would "gradually lessen and finally cease altogether," and Hawai'i's chief industry would be destroyed.[55]

As Spreckels' financial woes increased, Hawaiian Commercial & Sugar Company continued to expand, in 1889 gaining half interest in Waikapū Plantation. In 1890 Spreckels proposed a merger of Hawaiian Commercial & Sugar Company with Pā'ia and 'Haikū plantations, a deal that was rejected by Alexander and Baldwin because they were loath to take on Spreckels' debts. During this period, Spreckels also gained control of American Refinery, which had been processing most of Hawai'i's raw sugar, through a merger with his Western Sugar Refining Company. However, unable to service its debts, in 1893 Hawaiian Commercial & Sugar Company declared bankruptcy and was taken over by its bondholders.[56]

The 1890s were a period of consolidation and growth for Henry Baldwin and Samuel Alexander. In 1894 the Alexander & Baldwin agency was established in San Francisco, representing the 'Haikū and Pā'ia plantations. Baldwin was active in civic affairs, financing the Baldwin Home on Moloka'i to serve boys and men with Hansen's disease (leprosy), rebuilding his father's Waihe'e Church in Lahaina, and serving as a delegate to the 1894 convention that established the Republic of Hawai'i. In 1895 Henry Baldwin named his son Harry A. Baldwin the manager of 'Haikū Sugar Company.[57]

Castle & Cooke had long served as Alexander & Baldwin's agent, marketing their sugar and molasses and purchasing equipment and supplies for them in both Hawai'i and California. As a result, Alexander and Baldwin became indebted to Castle & Cooke. However, during the depression of 1891–1894, Castle & Cooke surrendered its debt and San Francisco affairs to J. D. Spreckels & Brothers, making the Spreckels' firm the San Francisco agent for the Pā'ia and 'Haikū plantations. This arrangement did not sit well with Alexander and Baldwin, who remembered too well their earlier dealings with J. D. Spreckels' father, Claus. In response, they raised the money needed to pay off J. D. Spreckels & Brothers and set up their own agency, Alexander & Baldwin, to represent their two Maui sugar plantations as well as their other properties, including Haleakalā and Honolua Ranches on Maui and Hawaiian Sugar Company on Kaua'i (later Olokele Sugar Company). In December 1894 Alexander & Baldwin opened an office in San Francisco, and within a month the company's first shipment of sugar left Hawai'i for the mainland. In 1898 Alexander & Baldwin purchased HC&S, finally gaining control of the plantation established by Claus Spreckels two decades earlier. Thus Alexander & Baldwin became one of the five major agencies serving the sugar industry, the others being Castle & Cooke, H. Hackfeld & Co. (later American Factors), C. Brewer and Company, and Theo H. Davies.[58]

As soon as the first East Maui ditches were completed, plantations realized that if they could store some of their ditch water, they could better manage its distribution. Maui plantations began to construct small reservoirs for temporary storage of water from the Hāmākua and Spreckels ditches. In 1882 *The Planters' Monthly* reported that "Makawao District, Maui, takes the lead in the matter of reservoirs for storing water," and went on to list "the reservoirs now in use on the plantations in that District: 'Haikū Plantation 2, Hāmākuapoko 6, Alexander & Baldwin 4, J. M. Alexander 1, Grove Ranch 3, Spreckels Plantation 2.'"[59]

A. H. Smith of Grove Ranch Plantation on Maui described the methods used to build two reservoirs. One reservoir had an 800-foot semicircular dam a little less than six feet tall and held 20 hours of irrigation water. Its $374 cost was primarily for 336 man-days of labor at 75 cents per worker per day.

The dams were made entirely of earth scraped up with scrapers drawn by oxen, the ground being first ploughed on each side of the line of the dam. I found the revolving scrapers the best for the purpose. We used two pairs of oxen to each scraper.

These reservoirs have proved to be of great value in saving all of the water, and enabling us to use it in the day time; they are of especial service when water is scarce.

Various expedients have been resorted to for resisting the action of the waves . . . stoning, planting with manienie grass [bermudagrass], barricading with logs, cane trash, [etc]. But the only effectual protection is found to be stoning, and even that has to be very carefully done where the reservoir is large and exposed to the trade winds.

I have found with the three reservoirs on this Plantation, that they all leaked more or less at first, but gradually became tight. Treading the reservoir with cattle or horses, when the ground is wet, helps very much.[60]

At the time of the Reciprocity Treaty, three plantations were operating near the mouth of the 'Īao Valley on the lower eastern slopes of West Maui: Wailuku Sugar Company, Bal & Adams, and Waikapū Sugar Company, all of which began operations during the American Civil War. These plantations were apparently irrigated with waters from the Wailuku River because G. W. Wilfong, who had become manager of Wailuku Sugar Company in 1868, noted that "diversion of water during dry seasons caused much quarreling which resulted in heavy law suits."[61]

In 1877 Wailuku Sugar Company acquired 420 acres of Bailey Brothers Plantation as well as the Bal & Adams Plantation. In 1894 Wailuku Sugar Company purchased Waihe'e Sugar Company to its north and Waikapū Sugar Company to its south. Located on the lower eastern slopes of West Maui above the isthmus, Wailuku Sugar Company needed to develop a ditch system that could tap the streams and aquifers of the northern and eastern slopes of West Maui. However, much of that water was already being captured by a ditch built in 1882 by Claus Spreckels, the Waihe'e Ditch (also known as the Spreckels Ditch on West Maui, not to be confused with Spreckels Ditch on East Maui). Spreckels had built this West Maui ditch to capture up to 60 mgd, beginning at Waihe'e stream, and carry it southward 15 miles to Kalua, where it emptied into HC&S's Wai'ale reservoir near Wailuku. This allowed Spreckels to irrigate his HC&S Plantation with water from both his 'Haikū Ditch on East Maui and his Waihe'e Ditch on West Maui.[62]

However, when Wailuku Sugar Company bought Waikapū Sugar Company in 1894, it began to contest Spreckels' right to the water in Waihe'e Ditch. When, in 1898, Spreckels sold HC&S to Alexander & Baldwin, Wailuku Sugar Company and HC&S entered into a complex agreement to develop and share West Maui water. The two plantations agreed to equally divide the water in the original Waihe'e Ditch. They would then construct a new Waihe'e Canal with 40 percent of the water going to HC&S and the remainder to Wailuku Sugar Company. Maintenance costs of the ditch and canal would be shared in proportion to the water received. HC&S would have a right to all surplus water from both ditches, and water from Waihe'e Ditch and Waihe'e Canal would be transferred to HC&S via Waihe'e Ditch only between 7:00 p.m. and 5:00 a.m. In addition, HC&S would receive all water from another new ditch, the South Waiehu Ditch, which took water from South Waiehu Stream that flowed to the sea just north of Wailuku.[63]

This complicated set of agreements provided both HC&S, now under the ownership of Alexander & Baldwin, and Wailuku Sugar Company with the ditch water needed to expand their cane acreage and sugar production after annexation of Hawai'i by the United States.

The rapid development of irrigation on Maui undoubtedly helped convince plantations on the other islands to invest in alternative means of water supply. In 1890 two new plantations on O'ahu—Kahuku Plantation Company and 'Ewa Plantation Company—were actively developing irrigation infrastructure. They had planted their first crops of cane, 700 and 650 acres, respectively, and *Thrum's Hawaiian Almanac and Annual* reported that "a number of artesian wells have been bored on these plantations, and others are yet to be sunk. At the 'Ewa Plantation, an extensive steam pumping plant has been erected to convey the water supply from six ten-inch connected wells to the fields first brought under cultivation. Another pumping station of double power and capacity is to be constructed to carry water to a still higher plane to serve another stretch of fine cane land." Similar development was occurring at Makaweli, on Kaua'i, where Alexander & Baldwin's new Hawaiian Sugar Company was developing "its massive water works of excavations, ditches, pipes, and flumes for the irrigation of its extensive tracts of cane land" even though its mill would "not be completed until the latter part of 1891."[64]

Forest Preservation

As sugar plantation owners began to develop irrigation, they realized that the Kingdom's forests were keys to slowing runoff, increasing infiltration of rainfall, protecting soils from erosion, and preventing their canals from filling with sediment. In 1876 the industry supported passage of an "Act for the Protection and Preservation of Woods and Forests" by the Kingdom of Hawai'i, emphasizing the need to maintain forest cover in order to assure more reliable stream flows.[65]

Even after passage of the act, *The Planters' Monthly* continued to denounce destruction of the islands' forests, recalling that in 1860 the mountains behind Honolulu "were covered with a dense forest growth . . . [and] the ground was a heavy mass of ferns and under-growth . . . [and] every gulch had its little stream gently trickling to the valley below where flowed a strong pure stream of sweet water." In contrast, in 1882 "the valleys and hillsides [were] almost wholly denuded of trees, the water-ways dry and rocky, except when filled with torrents of muddy water from the winter rains, great rifts torn from the mountain sides where land-slides have ripped away the very bowels of the hills." The water supply in 1882 was "impure and capricious as compared to that of twenty or thirty years ago. To say that all of this [had] no connection with the loss of forests is certainly a bold assertion."[66]

The presumed relationship between forests and rainfall induced James Makee to plant "many thousand Australian eucalyptus trees on the hillside [at 'Ulupalakua Plantation on leeward southwestern East Maui] in the hope of procuring a larger rainfall, so that the neighbourhood has quite an exotic appearance."[67]

Agricultural Diversification

The 1876 Reciprocity Treaty allowed for duty-free import into the United States of Hawaiian raw sugar, molasses, and other agricultural products, including arrowroot (*Tacca leontopetaloides,* called pia in Hawai'i), bananas, castor oil, nuts, undressed hides and skins, pulu (a tree fern fiber used to stuff mattresses), seeds, plants, shrubs, rice, tallow, and vegetables (fresh, dried, or preserved). In return, a wide variety of manufactured and agricultural products from the United States were allowed to be imported duty-free into Hawai'i. The elimination of tariffs strengthened the Kingdom's interest in exporting a greater variety of agricultural products, and by the 1890s, the Kingdom of Hawai'i was promoting diversification of its agriculture.[68] Of course, sugar remained the most important crop, with most of the production on larger plantations that ground their own cane and processed the juice into raw sugar. Some attempts had also been made to develop profit sharing systems in which small farmers cultivated cane and had it processed at the larger plantations.

Rice, which had been introduced in Hawai'i in 1858, was primarily produced by Chinese immigrants. The government reported that "neither the European nor the American can cultivate [rice] as laborers. It requires working in marshy land, and though on the islands it yields two crops a year, none but the Chinaman can raise it successfully." Though "dry-land or mountain rice" had been introduced, it was still unproven. Despite these constraints, by 1890 the rice industry farmed 7,420 acres and was producing an estimated two tons of rice per acre, almost all on O'ahu and Kaua'i. This made rice the second-most important crop in the islands, with acreage just over 10 percent that of sugarcane. The industry continued to grow, and as of 1907 rice was being grown on 9,400 acres, producing almost 21,000 tons on land that had previously been used by Hawaiians for taro production.[69]

Arabica coffee, an especially flavorful coffee indigenous to Ethiopia, was first introduced to Hawai'i by Don Francisco de Paula Marin in 1813. The first commercial attempt to grow coffee was near Honolulu in 1825. It failed rather quickly, but a second more successful attempt at commercial production was made on Kaua'i in 1835 at the same time sugarcane growing was initiated. By 1862 coffee had achieved second-place status following sugar but, in contrast to sugarcane, it was considered a crop for the small independent farmer. Coffee fell to third-place status following the rice boom. The Reciprocity Treaty, so beneficial to sugar, negatively affected the coffee industry in Hawai'i because coffee imports were allowed duty-free into Hawai'i, thus competing with island grown coffee. The Kaua'i coffee venture eventually failed due to an undetermined blight.

The coffee industry suffered additional setbacks in the 1860s, when drought, disease, insect infestations, and labor shortages caused nearly all plantations on the islands to close, except for those near Kona and Hāmākua. But strong economies in Europe and America during the 1890s caused coffee prices to rise, creating a boom for Kona coffee. In 1892 the German businessman Hermann Widemann introduced an improved Guatemalan coffee variety now known as Kona "typica."[70] In an 1896 pamphlet produced by the Hawaiian Department of Foreign Affairs, the commissioner of agriculture reported that "hundreds of thousands of [coffee] trees" had been planted in the early 1890s, and the government felt that it was the "crop of the future." It had the benefit of requiring less capital and land than sugar to be profitable, and "at the end of the fourth year the return from a 75-acre coffee plantation will much more than pay the running expenses." Profits of $8,000 to $10,000 per year could be anticipated after that. Because of these opportunities, in 1896 the Kingdom was encouraging farmers to establish small coffee plantations, initially living "a simple life" in which "the living must be close, and cash must be paid out for the necessary improvements" before "the crops begin to give returns." But "it only requires brains, a small capital and energy to realize such comfort and independence as cannot be realized in old [traditional coffee-growing] countries, in one-fourth the time."[71]

The mid-1890s also saw growth in fruit and vegetable exports from Hawai'i. By 1896 about 100,000 stalks of bananas worth over $100,000 were being exported each year. They were grown only on O'ahu, however, and expansion to the other islands would depend on securing markets. Pineapples with a value of nearly $9,000 were exported to San Francisco in 1895. Limes and oranges could also "be raised to perfection," and jellies for local consumption were being made from guava and poha or cape gooseberry.[72]

Around the same time, potatoes were being grown for local consumption "in the uplands of Hawai'i and Maui" by Portuguese and Norwegian farmers. Corn was also being grown, mainly as feed for cattle, horses, and mules on the sugar plantations. There was a large local market for pork, primarily for sale to the 15,000 Chinese in the Kingdom, and hogs were being fattened on corn and potatoes "in the Kula district of Maui." Other crops touted as having promise for diversified agriculture included temperate vegetables and fiber crops like ramie, sisal, and sansevieria. In addition, the English-born plant collector William H. Purvis had introduced macadamia nuts to Hawai'i in 1881, but the first commercial production did not begin until 1925 on O'ahu and 1929 on Hawai'i.[73]

Experiment Station, HSPA

Hawai'i's sugar planters had known for many years that fertilizer needed to be added to their soils in order to obtain good yields. As a result, in the first convention of the Planters' Labor and Supply Company in 1882, the company trustees were requested "to consider the advisability of employing a thoroughly competent chemist" to help the plantations diagnose and correct soil fertility problems. In 1892 the company's Committee on Fertilization renewed the call for "a chemist to serve the plantations and there is need for an experiment station with a laboratory." The call to establish an experiment station with a chemist and laboratory was renewed in 1894, with the proposal that the sugar companies be assessed five cents per ton of sugar for general expenses and five cents per ton for a laboratory and chemist. As a result, in April 1895 the Experiment Station was established when Dr. Walter Maxwell arrived in Honolulu from Louisiana to become its director. Maxwell had spent five years in Germany working on analysis of sugar beets and soils, four years in Washington where, as special agricultural expert, he set up a sugar beet station in Nebraska, and two years in Louisiana where he worked on sugarcane. A month later his assistant, Mr. J. T. Crowley, also from Louisiana, arrived, and the pair began to set up the chemistry laboratory. Within a few months Maxwell had visited all the islands and most of the plantations and had written reports of soils and fertilization. In

November 1895 the Planters' Labor and Supply Company became the Hawaiian Sugar Planters' Association with Maxwell the chairman of its Committees on Fertilization, Cultivation, and Manufacture. The HSPA's new emphasis on scientific agriculture began to emerge.[74]

Summary

Passage of the Reciprocity Treaty between the Kingdom of Hawai'i and the United States in 1876 eliminated import duties on Hawaiian sugar and other agricultural products, as well as manufactured goods from the United States. This increased the profitability of the islands' sugar companies and greatly stimulated investment in the industry. As a result, sugar production in the Hawaiian Islands increased twenty-fold, from 13,000 tons in 1876 to 260,000 tons in 1897. This growth, combined with a declining population of Hawaiians, required importation of thousands of laborers, especially from China, Portugal, and Japan. The sugar industry workforce lived on the plantations in company-owned "camps" or communities that were often organized by country of origin and provided housing, company stores, schools, churches, and health and recreational facilities. Much of the food required by the laborers was produced on the plantations. The paternalistic system of securing and managing plantation labor was effective, but many laborers left the industry as soon as their contracts allowed. Sugar growers struggled with chronic labor shortages, as well as periodic strikes and other forms of labor unrest.

Still, the industry was profitable and made major investments in company infrastructure, including mills, ports to handle exports, plantation communities, and railroads and "wet flumes" to transport cane to the mill. Extensive water harvesting systems composed of open ditches, tunnels, and large pipes were constructed to collect stream flows and shallow groundwater from high-rainfall windward areas and transfer it to cane fields in leeward areas with abundant sunshine but little rainfall. Maui sugar companies led the industry by developing major multi-plantation ditch systems on both East and West Maui. These were enhanced by constructing a network of small reservoirs to temporarily store ditch flows and reduce waste of irrigation water. Advances in field operations included the introduction of large steam-powered plows, partial replacement of hand hoeing with horse- and mule-drawn planters and cultivators, use of fertilizers such as guanos, and transportation of cane to the sugar factories by rail and wet flumes.

Major improvements were also made in milling and sugar-making equipment. Stimulated by the need to protect the islands' water resources, the sugar industry led major efforts to plant trees in areas where cattle and logging had caused major deforestation. Efforts to promote agricultural diversification met with limited success, but sugar remained by far the most important crop in the islands. Recognizing the need to cooperate to secure favorable legislation and an adequate plantation labor force, in 1882 the sugar companies formed the Planters' Labor and Supply Company, which in 1895 became the Hawaiian Sugar Planters' Association (HSPA). Further recognizing the need for a sound scientific basis for sugarcane production and milling, in 1895 the HSPA established an industry-financed Experiment Station.

3
Industry Growth and Labor Unrest—1898 to 1929

The period around the turn of the twentieth century was critical for Maui's sugar industry. Maui plantations led the industry in development of both surface and groundwater resources with continued development of the East and West Maui ditch systems and substantial investments in wells and reservoirs. Major improvements were made for in-field surface irrigation methods, use of fertilizers, cane harvesting and transportation methods, and factory operations.

But the forces that affected Maui's plantations had industry-wide impacts. Annexation of the Kingdom of Hawai'i by the United States in 1898 was supported by the sugar industry, in part because it assured that no new tariffs would be levied on Hawaiian sugar. But it also ended the industry's system of labor contracts to secure immigrant labor. New labor laws, the industry's chronic lack of labor, two significant strikes, and the general profitability of the industry led to substantial improvements in wages and the living conditions of the labor force on Hawai'i's sugar plantations. The industry's profitability also allowed it to greatly increase cane acreage, rapidly mechanize its operations, develop new and improved cane varieties, continue irrigation expansion, and increase the use of fertilizer.

Sugar Industry Development

In his enthusiasm for Hawai'i's becoming a U.S. territory, Henry Baldwin of Alexander & Baldwin Plantation held a large annexation luau at 'Haikū Plantation on East Maui. But the Organic Act of 1900, which established the rules under which the territory was organized, brought problems as well as opportunities. For example, it limited new corporate landholdings in Hawai'i to 1,000 acres (while exempting previously organized corporations). It also forbade importation of foreign labor, as well as labor contracts binding

Table 3.1. Hawaiian raw sugar production (in short tons, 2,000 pounds), measured every five years between 1900 and 1925

Year	Tons
1900	360,036
1905	429,213
1910	566,821
1915	592,763
1920	521,579
1925	787,246

Sources: U.S. Department of Agriculture, *Yearbooks of Agriculture,* 1924 and 1926, 1925 and 1927.

Table 3.2. Average New York sugar prices, in five-year increments, from 1894–1899 through 1920–1924

Years	Price (in cents per pound)
1894–1899	3.13
1900–1904	3.84
1905–1909	3.96
1910–1914	4.01
1915–1919	6.16
1920–1924	6.94

Source: Dorrance, *Sugar Islands,* 2000.

workers to the companies. As a result of industry optimism, as well as the act's limitations on plantation size, a number of new sugar companies were soon established. By 1912 the Hawaiian Islands had 53 sugar plantations, with 26 on Hawaiʻi, 10 on Kauaʻi, 10 on Oʻahu, and 7 on Maui.[1]

The early twentieth century was also a period of rapid technological progress. The U.S. Department of Commerce reported, "In the application of scientific methods and in the securing of results, the cane-sugar industry of the Hawaiian Islands will bear comparison with any agricultural or mechanical industry in the United States." The industry dominated Hawaiian agriculture. With more than 45,000 persons employed on sugar plantations, sugar accounted for about nine-tenths of the agricultural production of the islands.[2]

Sugar production increased throughout the first quarter of the twentieth century, more than doubling by 1925, primarily because the price of sugar more than doubled between the late 1890s and the early 1920s (see Tables 3.1 and 3.2).

Plowing, Planting, and Cultivation

During the first decade of the twentieth century, field work for a typical irrigated sugarcane crop began with the land being "thoroughly ploughed up by steam ploughs." This often meant that the field was first plowed, then cross plowed, then harrowed. Planting and irrigation furrows were then laid out by survey teams on the contour about 5 feet apart. These furrows were plowed with a "double-decked moldboard plow" to a depth of 24–32 inches. Lateral ditches to carry the water from the main ditches to the furrows were cut at intervals of about 50 feet. The seed cane pieces were then laid end to end in the bottom of the furrow and covered with a few inches of soil from either side. Irrigation water was, as soon as possible, brought from the main ditches down the lateral field ditches and turned into the furrows, running down the furrow to the next lateral ditch about 50 feet away. Immediately after planting, the cane was irrigated every two days. After the cane leaves emerged the interval between irrigations was increased according to the experience of the manager.[3]

Mechanization of sugarcane field operations increased rapidly after 1900. On some unirrigated plantations, two-row planters opened deep furrows, dropped seed pieces into them, covered the seed pieces with soil, and firmed the soil over the furrow—all in one pass—greatly reducing the labor needed for planting.[4] Four- and six-horse teams used for plowing were rapidly replaced by pairs of steam engines.

> In the preparation of the land, steam plows are used, operated by powerful engines that work in pairs, on each side of the field. Gang plows attached to wire cables that are wound upon drums on the engines are pulled back and forth across the fields. Each gang plow requires a minimum of three men in its operation, exclusive of the enginemen and the water-supply men. These plows turn furrows from 18 to 36 inches deep, breaking and pulverizing the soil finely. . . . Mule plowing is done in Hawai'i on portions of fields in which steam plows cannot be operated and upon land where the latter are not available.[5]

Prior to 1920 weeds were usually controlled by hand hoeing on irrigated plantations and with disc cultivators pulled by mules on unirrigated plantations. But chemical weed control was also beginning. "On a large unirrigated plantation having a heavy rainfall and rank weed growth and managed by a scientist, the destruction of weeds and grass [was]

Figure 3.1. Filling horse-drawn herbicide sprayer from tank on wagon. Photo courtesy of HARC

accomplished by the use of arsenic sprays between the rows of cane. The sprays were applied both with backpack sprayers and from tanks of arsenic solution mounted on sleds pulled between the cane rows by animals (see Figure 3.1). This method [was] an economical substitute for a large amount of hand hoeing and implement cultivation."[6]

Ratoon crops were cultivated somewhat differently than plant crops. After harvest, "part of the earth forming the hills between the furrows [was] turned over into the furrows by plows," covering the "short stubs of the cane with 4 to 5 inches of earth." Soon thereafter, the cane was irrigated, and "in a short time new sprouts [would] spring from the buds on the old stalks of the cane" and the ratoon crop was under way. "Hilling up" referred to loosening the soil between the furrows with a small plow, then rebuilding the hill between the furrows, usually with a double-moldboard plow. This was thought to aerate the soil and stimulate growth of the ratoons.[7]

Horses pulled fertilizer spreaders, and donkeys carried cane trash on their backs to feed livestock. By World War I, gasoline-powered tractors were beginning to compete with steam tractors, and on May 26, 1916, the *Maui News* reported a "demonstration of a 90 horsepower gasoline tractor of the caterpillar type in plantation plowing" at HC&S Plantation.[8]

Around the turn of the twentieth century, sugar growers often removed dead leaf blades and leaf sheaths from the growing cane (called stripping) in order to harden the stalks to resist cane borers, allow light to reach young secondary stalks, make harvest easier, and reduce the danger of accidental fire. The fields were normally "stripped" once or even twice by hand between January and June. The second stripping was done "when work [was] slack but labor [had] to be employed." Stripping was "favored on unirrigated plantations, but on irrigated plantations in the warmer and dryer belts it [was] considered by [many] managers an unnecessary expense and detrimental to the cane." Planter James Girvin indicated that stripping was "one of the hardest jobs on the plantation, not only on account of the extreme heat in the center of a field, but from the sharp serrated leaves of the cane whose edges cut like a saw and knife combined."[9]

Irrigation

Early in the century, in-field irrigation management was still an art.

The Department of Commerce[10] reported that "in Hawai'i . . . the best results are obtained when the young cane received 0.5 inch per week; less favorable results are obtained when the water supplied was 1 inch per week, and when the furrows are filled with water the cane comes up yellow and sickly. As the cane comes away it requires about 1 inch weekly up to three or four months, after which 1.5 inches are necessary until the crop is in full vigor, when 3 inches and never more are required." As the cane grew, dead leaves and leaf sheaths fell into the furrows, and during the second year of growth the stalks became too tall to stand and fell over, or lodged. The growing points of stems that had lodged responded to gravity and turned upward, only to lodge further a few months later. By harvest time the field was a tangled mat of vertical and horizontal stems. The dead leaves and lodged cane that formed this mat slowed water movement in the irrigation furrows, and during the second summer of growth it became more difficult for irrigators to move water to the ends of the rows. Irrigation was normally stopped three to five months before harvest to dry the field off, to aid in harvesting, and to increase sugar content in a process called ripening. For irrigated plantations in the islands, irrigation standards in the early decades of the century were:[11]

- Every 100 acres of cane required 1 million gallons a day (mgd) of irrigation water.
- One thousand pounds of irrigation water was needed to produce one pound of sugar.

- During a period of growth of about 17 months, the total water supplied to the crop averaged about 100 inches.

Irrigation was the most costly operation on the 24 irrigated plantations audited by the Department of Commerce for its 1917 publication (Table 3.3).[12] Three of the plantations had irrigation costs exceeding $100 per acre, but the average cost was $68 per acre or $11 per ton of sugar. As a result of the importance of irrigation, in 1919 lecturer and author R. M. Allen provided a short course for plantation irrigation personnel. In it he taught irrigators how to reduce irrigation water losses, how much land could be irrigated with a given amount of ditch water, and how to calculate the amount of water that should be let into the ditch to irrigate a certain amount of land. Other topics included how to build weirs and measure ditch flow rates.[13] He noted that in 1913–1914 Hawai'i's wholly and partially irrigated plantations spent $15 million on field operations, of which $4.5 million was for irrigation (60 percent for pumping and 30 percent for labor), $2.7 million for fertilizer, and $2.4 million for harvesting, and the remainder for planting, cultivating, and other activities. Producing one ton of cane required 7 to 14 acre-feet of water for each acre of land that produced a full two-year crop.[14]

Losses of irrigation water were often substantial. Seepage losses in main unlined plantation irrigation ditches ranged from near zero for some canals in compact soils to 50 percent per mile in some crossing porous soils; losses of about 25 percent per mile were common. Lining canals whenever possible, beginning with the leakiest sections, could reduce these losses. Leakage from in-field ditches that conveyed water to cane rows, especially level ditches, could be reduced by increasing the speed with which fields were irrigated, thereby reducing the time that water flowed in the in-field ditches.

Table 3.3. Average costs per acre for producing sugarcane, 1913–1914

Item	Irrigated	Unirrigated
Planting	$19.17	$13.84
Cultivating	$24.15	$46.52
Irrigating	$67.91	—
Fertilizing	$40.37	$48.82
Harvesting	$36.03	$37.93

Source: U.S. Department of Commerce, *The Cane Sugar Industry,* 1917.

Leaky gates could be fixed, and "general careless, poor methods of irrigation" could be improved by closer supervision of irrigators. Despite minimizing leakage, application of excessive amounts of irrigation water could cause water to move below the root zone and be lost to the crop. For example, on the Waipi'o field station of the Hawaiian Sugar Planters' Association (HSPA) Experiment Station, 47 percent of a 6-inch irrigation and 65 percent of a 9-inch irrigation were lost below the root zone. This type of loss was doubly serious when fertilizer nitrogen was applied in the irrigation water because it could also be lost below the root zone. Irrigators were warned that overirrigation was of particular concern in the first few months of plant crops before root systems had reached their maximum depths. Finally, overirrigation could cause waterlogging in some soils and could exacerbate soil salinity problems in others.[15]

Largely as a result of the great cost of irrigation, by 1926 plantations reported a wide variety of activities focused on improving irrigation water use. These included careful measurement of water at strategic locations in canals and at the head of each field. "Regulated irrigation experiments" were under way at several plantations to measure amounts of water applied to sections of the field. Experiments were conducted in the winter, summer, and for ripening cane with crop response measured in terms of cane growth rates and, in ripening experiments, sucrose concentrations. Experiments with sprinkler irrigation were also being conducted in 1926, but these failed and sprinkler irrigation did not emerge as a viable method until after World War II.[16]

Plantations also experimented with a number of in-field irrigation methods. The early irrigators at Pioneer Mill Company worked in teams of two. One turned water from a ditch into the upper end of a long furrow while his coworker waited at the lower end. When the water reached the lower end he would call out *piha*, meaning "full," and the ditch water would be diverted into the next furrow. This method is reported to have lost favor with Pioneer Mill management one night when the manager was kept awake by shouts of *piha* in the field near his house. The next morning the irrigator lost his helper; it is said that "many years passed before a further labor reduction of 50 percent was possible in irrigation."[17]

The "contour" irrigation method was used by most plantations in the early twentieth century. It involved planting cane in furrows about 5 feet apart, laid out on the contour across the slope. In-field ditches called "watercourses" were laid out about 35 feet apart running down the slope. Water was turned into these watercourses from larger ditches located 200–300 feet apart on the contours. The water ran swiftly down the watercourse until the irrigator turned it into (and quickly filled) one of the short-level furrows. This furrow was then closed, and the water in it was allowed to infiltrate while the water from the watercourse was turned into the next furrow down the slope. This contour system was effective and widely used, but it could only be used in soils that could resist the erosive effects of the water moving rapidly down the watercourse.[18]

Another irrigation method came into use in the 1920s for slightly sloping fields, initially at 'Ewa Plantation Company on O'ahu and Kīlauea Sugar Plantation Company on Kaua'i. Known at 'Ewa as the "no watercourse system," water was taken from supply ditches running on the contour and introduced into long furrows running down the slight slope. Rather than simply filling a short furrow, the water infiltrated as it made its way down the long furrow. By the time it reached the far end, the upper end of the furrow could be closed, and any excess water could flow out the lower end into the next supply ditch.[19]

On extremely flat fields, like some at 'Ewa Plantation Company, it was difficult to obtain enough flow to force the water down the furrow. Beginning in 1925, large areas were enclosed in substantial levees, and the whole area was flooded, much like a rice paddy or taro field, except that the field was not left submerged for more than a few hours.[20]

A final irrigation method came into use in the late 1920s on fields with slight but definite slopes. In these areas, long strips of land, usually four cane rows wide, were laid out down a long and gentle slope and surrounded by berms, or borders. These strips were level from side to side between the berms; therefore, when water was introduced at the upper end of the strip, it spread laterally across the four rows of cane then moved down the slope in a thin sheet. This "border" or "level border" method was very efficient in terms of labor, but its use was limited to gently sloping areas where the level borders or strips could be installed.[21]

Fertilization

During the early twentieth century chemical fertilizers were often still referred to as "manures." H. C. Prinsen Geerligs,[22] in his summary of sugar industries worldwide, reported that in Hawai'i "it seem[ed] to take a heavier manure each time to keep the cane production up to the same standard." Consequently, the industry relied increasingly on commercial fertilizers to replace nitrogen, potassium chloride (potash), and phosphate taken up by the crop and lost when the crop was burned or hauled to the mill with the cane. In addition, plantation managers were aware that nitrate moves with water in the soil, and some were concerned that excessive irrigation was leaching plant nutrients below the root system, causing them to be unavailable to the crop. The most common commercial fertilizer products included ammonium sulfate, Chilean sodium nitrate (saltpeter), potash, and phosphate. Prinsen Geerligs estimated that the industry spent $4.65 per ton of sugar or $22.20 per acre of cane harvested to purchase fertilizer. In addition to commercial fertilizers, filter cake and manure from the plantations' stables were used to improve soil fertility.[23] For the 1913–1914 crop, the average total cost of all these materials, including their application, was about $40 for irrigated and $49 for unirrigated plantations. In general, 600–1,000 pounds per acre of a mixed fertilizer (with 7–10 percent each of nitrogen, phosphate, and potash) were applied during the first growing season, and up to 400 pounds per acre of sodium nitrate (or sometimes ammonium sulfate) were applied in the second year.

Filter cake from factory filter presses was valued as a fertilizer because it was rich in organic matter, containing both organic nitrogen and phosphorus. Though each plantation managed it differently, filter cake was often mixed with ash from the mill furnaces (which was rich in potash) and/or stable manure (rich in nitrogen, phosphorus, and organic matter) and applied to the fields. Some plantations composted the cake with ash and other materials before applying it to fields. Some hauled it to the fields in wagons. Others bagged it and hauled it with mules. Still others transferred the filter cake to the fields in railcars. Some applied the material only to areas of fields where cane grew poorly, while others applied it to broad field areas using manure spreaders. Some incorporated the materials into the soil with steam plows; others placed it in planting furrows by hand. Some irrigated plantations dumped the materials into irrigation ditches and let the irrigation water carry the nutrients to the crop (see Figure 3.2). Regardless of the methods used, most plantations valued filter cake, furnace ash, and stable manure as important fertilizer materials.[24]

Potassium is required by sugarcane in greater amounts than any other nutrient, even nitrogen. About 550 pounds of potash (K_2O) was removed from the field with every 100 tons of cane harvested. A substantial amount of that potash ended up in molasses, which normally contained 3–6 percent potash. Experiments with responses to fertilizer potassium began to be

Figure 3.2. Dissolving and applying granular fertilizer in
irrigation water. Photo courtesy of HARC

reported in the 1920s on unirrigated plantations on the island of Hawaiʻi.[25]
In addition, the 1920s reports from Java and Mauritius suggested that molasses could be a good source of potassium if applied to cane soils. Further-
more, Pioneer Mill Company had conducted experiments on the use of molasses as a source of potassium after "a railroad tank car [filled with molasses]
was accidently upset in a cane field, with the result that the cane made an
unusual growth in the small area so saturated." When, in the late 1920s
molasses prices in California, the principal market for plantation molasses,
dropped so low that they hardly covered the cost of shipping, HC&S began to experiment with applying molasses in irrigation water to selected cane
fields, using railroad tank cars to bring the molasses to convenient locations
beside irrigation ditches. The entire daily output of molasses by the factory,
about 70 tons, could be applied by one man managing the flow of molasses

into the irrigation water and three men managing the irrigation of about 3.5 acres per day. The application rate was about 20 tons of molasses per acre. At 4.5 percent potash, this represented an application of about 1,800 pounds of potash per acre, several times the amount that would normally be applied to a cane crop, even one grown on potassium-deficient soils. Some of the level ditch sections of field appeared to respond positively to these applications of molasses.[26]

The importance of healthy root systems to absorb water and nutrients was also recognized, and in 1926 H. Atherton Lee, a pathologist at the HSPA Experiment Station, conducted studies of cane root development. At Wailuku Sugar Company and HC&S, the root distributions of several cane varieties were studied, as were differences in the rooting of cane planted in the bottom of the furrow versus cane grown on ridges between furrows. In almost all cases more than 50 percent of the roots, by weight, were found "in the topmost 8 inches of soil and more than 75 percent of the roots [were] above the 16-inch level." In all, more than 85 percent of the roots were found in the top 24 inches, leading to the logical conclusion "that the cane does most of its feeding in the uppermost 24 inches of soil."[27]

Some plantations, especially in unirrigated areas, did not burn their cane prior to harvest because the fields were sometimes too wet to burn. The cane trash, made up primarily of leaves and leaf sheaths, was known to have "great value for enriching the soil," because each ton of trash contributed 1.95 pounds of nitrogen and "large quantities of humus, which increase[d] the water-holding capacity of the soil for periods of drought."[28]

Harvesting

On most irrigated plantations, the cane was burned just prior to harvest to remove leaves and reduce cutting and loading costs. The cane harvest continued to rely on hand labor, with cane cut close to the ground and topped with cane knives. The long stalks were stripped of any leaves that might remain after burning. The stalks, which could be up to 25 feet long, were then cut into convenient lengths and thrown into piles for loading.

> The loaders follow the cutters. It is the custom for Japanese loaders to bring their wives to the fields. The wife lays a strap on the ground and piles 80 to 100 pounds of cane stalks on the strap. The man then slings the bundle on his shoulder and carries it to an inclined runway to a railway car. From the

permanent track of the company railroad a portable track is laid into the fields so that the cars will be at convenient distances from the loaders.[29]

During the 1913–1914 season the average contract sugarcane cutter harvested 7.2 tons of cane per day and earned an average of $1.11 per day. The average loader handled 6.7 tons of cane and earned $1.21 per day.[30]

If the cane had not been burned prior to harvest, the trash might be removed, burned, or cut into smaller pieces by large cane trash choppers with 10-foot cutter wheels that were pulled across the field by cables attached to steam engines, chopping the trash to facilitate its incorporation by subsequent plowing. The furrows were then "banked" (re-formed by throwing soil away from the rows with plows), fertilizer and irrigation were applied, and ratoon growth began. Fields harvested early, in November through January, would be old enough to flower profusely beginning the following November. Therefore, they were harvested at an age of approximately one year and were called "short ratoons." Fields harvested later than January would not be harvested the following winter. Known as "long ratoons," they would be left to grow until the next harvest season, about 22 to 24 months after planting. Of course, short ratoons had much lower yields than long ratoons. If a cane field had been harvested very late in the harvest season, the young ratoon crop might be "cut back" to ground level in June or July to set back its development and prevent it from flowering in November and December. It would be left to grow until the following harvest season, when it would flower and be harvested as 17- or 18-month-old cane. Many plantations allowed their fields to produce ratoons for six to ten years before they were plowed and replanted. Lahaina, Rose Bamboo, and Yellow Caledonia were the predominant varieties, with Maui and Oʻahu using Lahaina exclusively.[31]

Transportation to the Factory

Prior to 1900, cane wagons, sometimes with solid wooden wheels, were loaded by hand or by dragline and were pulled by mules to the factory. By the beginning of the twentieth century, plantations were mechanizing, and because of differences in topography and water availability, they were using a variety of methods to transport cane from the field to the factory. The most common means of cane transport was by "narrow-gauge railways, of which 840 miles of fixed and moveable railroad exist[ed] in the different islands, together with 120 engines and 8,500 wagons."[32] For example, one (unnamed) plantation reported:

Nearly all the cane is transported to the mill by rail. Our main line is of the standard 3-foot gage, and we use portable track, 20 pounds to the yard, sections 16 feet long . . . and galvanized steel ties. This track is satisfactory in every respect. The steepest grades where we use portable track is about 6 per cent. The cost of handling cane in cars on portable track is materially increased on fields where the grades are steep. More teams and teamsters are necessary to haul the empty cars into the fields, and more brakemen are needed in running the cars out of the fields by gravity onto the main lines of track. Cars are more apt to jump the track on steep grades, resulting in broken cars and making it necessary to reload the cane.

Another plantation used two steam-powered engines equipped with special large drums carrying 3,000 feet of steel wire rope almost an inch thick. These two engines could pull two large cane cars, each loaded with 6 tons of cane, up permanent rail lines running at slopes of 5–35 percent from lower fields to the main plantation line. Wailuku Sugar Company laid its portable track into the fields so that mules could pull the empty cars up the grade into the fields. After loading, the full cars moved by gravity back to the main railroad line.[33]

In addition to plantation railroads, 370 miles of flumes carried cane from the fields to the mill or to transfer stations on plantation rail lines. Flumes were mostly used in areas with abundant surface water that could reliably be diverted from streams. Prinsen Geerligs described the flumes as "shallow and boarded on each side, slightly sloping down from the fields to the factory, and ending in a kind of grate close to the mills." When the cane was harvested in the fields, it was "cut in pieces of 4 to 5 ft. long, and thrown into the gutter [flume], so that a constant stream of cut cane keeps flowing to the factory."[34]

Where ample water was available, modified flumes were sometimes used for temporary storage of cane awaiting milling. In 1913–1914 one plantation reported:

All our cane transportation is by flume. We have steadily increased our storage flumes in length and strength. For the past year we have used a 2 [inch] by 14 [inch] planking in all storage flumes. We have also converted about 3½ miles of our main flumes from the ordinary beveled flume to what we call a semistorage flume, thus eliminating 4 watchmen on a shift. In addition to the saving of labor, it has run the entire season without a flume jam, which is a great advantage in the steady feeding of our mill.[35]

Wailuku Sugar Company combined flumes with narrow gauge rail to move its cane to the factory:

> Fluming direct to the mill is no longer practiced here. This has been done to . . . enable us to distribute our water to better advantage for use in the fields, and . . . to allow us to weigh all our cane before grinding it. From one-half to two-thirds of our cane is harvested by the use of portable track in connection with our main railroad line, and the balance of the crop is flumed to railroad cars, the main flume leading to loading stations along the railroad line, which are located so that the water coming from these flumes can be used again for irrigation. There are six of these loading stations. . . . A new fluming station we are building this year is located at the mill supply reservoir. The water from this cane flume will be used at the mill after the cane has been flumed to the cars. This water, after passing through the mill, will again be used for irrigation purposes on the fields below"[36]

Many fields were too steep for access by rail and lacked the reliable water resources needed for flumes. As a result, plantations, principally on windward Hawai'i, developed a total of 40 miles of "suspended aerial ropeways" or tramways. These tramways usually consisted of wire cables strung on supports above the fields and extending down the slope to the factory or to a transfer station. In most cases bundles of cane were hung from a "trolley" that was attached to the tramway cable. When the trolley was released, gravity took it down the cable to the factory or transfer station, where the cane was removed and loaded onto a railcar or into a flume.[37]

In some cases tramways were used to bring cane up from lower elevations. On one plantation, "cane grown in gulches and on hillsides [was] in some cases hauled up on a standing wire rope, bundles slung under a trolley and hauled up by a gasoline engine," after which it was flumed down to the mill.[38]

Factory Operations

During the grinding season, factories typically operated 24 hours a day, 6 days a week, with mill employees working 12-hour days in six-hour shifts. Quality control was assured by a mill laboratory staffed with chemists who routinely sampled and analyzed cane, juice, syrup, sugar, molasses, waste water, filter press cake, and bagasse.[39]

When cane reached the factory via the plantation railroad, the railcars were weighed to determine the amount of cane being delivered. A mechanical unloader with long fingers then reached into the railcar, lifted the cane, and placed it on a slowly moving carrier that took it to the crusher. In the more modern factories, the first step in the milling process was to pass the cane through a series of rotating knives to chop it into small pieces. It then passed through two large steel rollers that crushed it, expressing the first and purest juice into a juice tank below, and producing an even mat of cane. The mat then moved through a set of cane cutters or teeth that shredded the cane before it passed through three or four (or up to six) sets of 3-roller mills. Arranged in triangles with two rollers below and one above, the mills used hydraulic pressure or springs to apply great pressure to the mat of shredded cane, expressing most of the remaining juice. Between the sets of rollers, hot water was sprayed onto the mat of cane to wash any remaining sugar into the juice tanks below. This process was known as maceration and left an average of only 4–5 percent of the original sugar in the fibrous mat of bagasse that exited the last of the roller mills.[40]

The size of the roller mills determined the rate at which the factory could crush its cane. For example, a crusher 30×60 inches followed by a 9-roller mill (three 3-roller mills) could crush 20 tons of cane an hour with 93.5 percent sugar extraction. A crusher 34×78 inches with a 15-roller mill (five 3-roller mills) could crush 60 tons of cane an hour with 94.5 percent extraction. After the mat of cane passed through the last set of mills, the remaining bagasse was dry enough to burn, and it was moved on a conveyer to the boiler, where it was used as fuel to power the mill, or to a storage area for later use.[41]

In addition to sugar and water, the expressed juice contained a quantity of gums, waxes, salts, soil, and other impurities. In order to remove these impurities from the juice (a process known as "clarifying"), the juice was pumped from the tanks to the top floor of the factory, where lime was added and the mixture was then heated to near boiling. The juice was then moved to large settling tanks, where after some time most of the solids settled out, and the partially clarified liquid was moved to cylindrical iron filter tanks (packed with wood fiber, bagasse, and/or sand) that removed practically all the remaining solids.[42]

The solids and unclarified liquids, or "mud," left in the settling tanks were forced through filter presses that used cloth filters to retain the solids while letting the liquids pass through. The liquid was returned to the clarifiers while the solids were scraped off the filters and taken to storage areas, where they could be collected until they were returned to the cane fields as an organic fertilizer and soil amendment.[43]

The clarified juice moved to multiple-effect evaporators, where steam and vacuum pump action were used to concentrate the sugars. Each evaporator contained chambers heated by steam that transferred heat to chambers filled with juice. The heat from the steam chambers boiled the juice, and the steam generated by the boiling juice moved through a pipe to the steam chamber of the next evaporator. In the first evaporator, the juice was at atmospheric pressure, and the steam temperature was well above 212°F. But the temperature of the steam in the second evaporator stage was less than in the first. In order to boil the juice in the second stage, vacuum pumps were used to decrease the air pressure, allowing the juice to boil at a lower temperature. As the juice temperature decreased in each successive stage, the air pressure above the juice also had to be reduced to induce it to boil. By the time it left the last evaporator, the juice had become syrup, with only 25–35 percent of the water it had when it left the clarifier. In the first decade of the twentieth century, island mills typically had three- or four-stage evaporators,[44] but by 1913–1914 four- and even six-stage units were the norm.[45]

The concentrated syrup from the multiple-effect evaporator was transferred to a vacuum pan, a large cylindrical vessel in which the syrup was heated with steam and a vacuum was maintained with pumps. As the water boiled away, the operator—known as the "sugar boiler"—watched for sugar crystals to form in the "mother liquor" containing dissolved sugar and impurities. As more and more sugar crystals formed, the mixture, known as "massecuite," a French term meaning "cooked mass," became thicker. When the crystals reached the right size, the sugar boiler made a "strike," releasing the vacuum and opening a valve in the bottom of the vacuum pan to let the massecuite flow to a mixing tank, where it was continually agitated to prevent the crystals from settling to the bottom and hardening.[46] The massecuite was then passed to perforated brass baskets inside centrifugal machines, vertical cylinders about 40 inches in diameter and 30 inches tall. These centrifugals were spun up to about 1,000 revolutions per minute by electric or steam motors, forcing the mother liquor out of the brass basket and leaving nearly dry raw sugar within. When the centrifuges stopped the raw sugar crystals were removed to bins to await packing in sacks of about 125 pounds each and then readied for export. The final molasses produced by the centrifugal was sold as cattle feed, used as a supplementary fuel for the furnaces, or applied to the fields as a source of potash.[47]

In his review of the world's sugar industries, Prinsen Geerligs[48] praised Hawai'i's sugar factories, noting, "The extraction results thus arrived at are never met with elsewhere, nor have they been surpassed. With so high a sugar content as the Hawaiian cane possesses, the sugar content of the bagasse is reduced to 3 percent." He attributed this extraordinary result to the fact that in Hawai'i, unlike Cuba, Java, and other tropical areas, "the cane in the field keeps good condition long after it is full-grown, while no early rainy season is to be feared, so that there is not a single reason for hurrying over the grinding, but everybody can work carefully, and try to get as much sugar as possible out of the canes." Prinsen Geerligs also reported that "owing to the high sugar content of the cane, as well as to the high purity of the juice and the high juice extraction by the mill, the sugar yield on 100 [units of] cane is extremely high in Hawai'i." During the 1913–1914 season, the average extraction in Hawaiian mills was 95.46 percent, with some mills reaching 98 percent. An additional advantage noted by Prinsen Geerligs was that most of the mills in Hawai'i were relatively new, and rarely needed fuel other than bagasse and waste molasses to generate the steam needed to power their operation.[49]

The first two decades of the twentieth century saw radical changes in Hawaiian sugar mills. Efforts to improve preparation of the cane led to the development of more efficient knives and shredders, and "by the end of this period knives had been installed at most of the mills and shredders at about one-third of them." The invention of "Messchaert grooves" in the roller mills let juice run out of the cane more freely as it was pressed between the rollers, reducing reabsorption of the juice by the cane fiber as it moved out from between the rollers. In addition, hydraulic pressure was being used to increase pressure on the cane, serving not only to allow the mills to open and avoid damage when pieces of metal trash entered them, but also to apply the pressure needed to rupture cells and release more of the sugar contained in the cane. Finally, "maceration," the addition of hot water to the mat of cane as it passed through the mills, became a "fairly common practice." The percentage of sugar extracted by the milling process increased each year, and in 1920 the average extraction of sugar from the cane reached 97.45 percent.[50]

During the 1920s there was little change in the fundamentals of crushing and milling of the cane, though heavier pressures and more powerful motors were usually implemented when it was necessary to replace aging equipment. The main change that occurred was a 30 percent increase in the tonnage of cane milled per hour. This was in response to an approximately 75 percent increase in cane production, which also required many plantations to extend the grinding season.[51]

Prior to the 1920s the industry had no great incentive to develop "elaborate schemes for steam economy [because] the supply of bagasse was usually sufficient" to meet the needs of the mills. However, as irrigated plantations used more energy for pumping, and opportunities to sell electric power to municipal grids increased, greater emphasis was placed on fuel economy For example, Pioneer Mill Company began using excess steam from its quadruple effect evaporator to preheat juice.[52] Also during the 1920s work by the HSPA Experiment Station revealed that clarification of juice could be enhanced by adding more lime to increase the alkalinity of the juice. In addition, new continuous Dorr clarifiers that were "unquestionably superior in every way" became available and were quickly adopted by the industry. Factories also experimented with different methods of extracting sugars from the mud that settled during clarification. These included completely eliminating the filter press and returning settlings to the milling plant where the solids could be discarded with the bagasse. Another approach was to use centrifugal separators to extract soluble sugars from the mud, but the extra water required to extract these sugars required more steam for its evaporation. By the 1920s the best new filtration technology appeared to be the new labor-saving Oliver filters with fine metal screens rather than cloth filters.[53]

Despite its excellent environment and infrastructure for sugar production, Hawai'i had high labor and transportation costs that made the industry dependent on U.S. government protection provided by tariffs on foreign sugar. Prinsen Geerligs[54] concluded that, on average, Hawai'i produced sugar for $55 per ton and sold it for $80 per ton. But $33 of the $80 per ton resulted from the U.S. government tariff on foreign sugar, which substantially raised the price to the U.S. consumer. Without this protection, the Hawaiian sugar industry could not have survived as it existed in 1910.

In 1913–1914 the average cost of growing and transporting cane to the mill was $37.56 per ton of sugar. The cost of milling added another $5.85 per ton, most of which was for labor, sugar containers, and repairs of machinery and buildings. Shipping to the United States added $8.66 per ton, for an average total cost (excluding depreciation) of $52.07 per ton. The U.S. Department of Commerce estimated that "the gross proceeds of the sugar crop of 1914 delivered in the United States were $68.37 per ton." While still impressive, this was the lowest price the industry had received in several years, a substantial decrease from the over $80 it received in 1910–1912.[55]

As a result of the industry's profitability, Hawai'i's production of raw sugar continued to grow steadily between annexation in 1898 and the onset of the Great Depression in 1929, more than doubling between 1900 (360,036 tons) and 1925 (787,246 tons) (see Table 3.1). The continued profitability of the industry was largely the result of increasing sugar prices, which more than doubled from 3.13 cents per pound in the years between 1894 and 1899, to 4.01 cents per pound in 1910–1914, and 6.94 cents per pound in 1920–1924 (Table. 3.2).[56]

Labor

After annexation in 1898, the terms of the Chinese Exclusion Act of 1882 were applied to Hawai'i, preventing immigration of additional Chinese to the islands. Soon thereafter, the Organic Act of 1900 was passed, defining the terms under which Hawai'i became a territory, abolishing labor contracts, and greatly reducing the plantations' hold on labor. And so the sugarcane industry was forced to revise its employment practices and look beyond China for labor. Japan provided the largest source of new labor in the first years following annexation. Almost 20,000 Japanese recruits arrived in 1899, and by 1900 about 60,000 Japanese were in the Islands.[57]

In 1903 the first Koreans arrived in Hawai'i, and six years later over 2,000 Spaniards arrived from Malaga, a traditional sugarcane-growing region. In addition, the Hawaiian Sugar Planters' Association entered into an agreement with the government of the Philippines, bringing the first 200 laborers from those islands in 1906–1907.

In the months before the Organic Act of 1900 became official, Japanese workers at Pioneer Mill Company on Maui, angered by the accidental deaths of three workers in the factory, refused to work, demanding compensation to the families of those killed. The strike soon spread to neighboring Olowalu Company, where Japanese workers demanded that management fire most overseers, reduce the hours in the workday, and change the system of "docking" workers' pay for misconduct. Japanese workers on the Spreckelsville plantation (Pā'ia Plantation Company) also went on strike, clashing with police and overseers, who used whips to force them back to work before management agreed to terminate their labor contracts. By the end of 1900 more than 8,000 workers—mostly Japanese—on more than 20 plantations had struck to demand concessions from management, including wage increases, shorter workdays, and better treatment by overseers. In many cases the workers' demands were at least partly met (see Table 3.4).[58]

The widespread strikes of 1900 emboldened plantation workers, especially those from Japan, and in 1905 and 1906 isolated strikes of Japanese workers

Table 3.4. Labor strikes on Hawaiian sugar plantations after 1898

Year	Plantations, island	Ethnic group	Issues	Outcomes
1900	Over 20 plantations	Mostly Japanese	Labor contracts, worker deaths	Workers won numerous concessions, including higher wages, shorter workday, dismissal of some overseers
1905	Pioneer Mill, Maui	Japanese	Beating of worker	Riot, one killed, concessions made by management
1906	Oʻahu Sugar Co.	Japanese	Higher wages	Concessions made by management
1909	Oʻahu plantations	Japanese	Higher wages, end ethnic wage discrimination	Promoted by Higher Wage Association, strikers fired and evicted, strike lasted 4 months, ethnic discrimination ended, wages increased, HSPA increased Filipino immigration
1920	Oʻahu plantations	Japanese & Filipino	Higher wages, end bonus system, 8-hour day, insurance, maternity leave	Strikebreakers hired, strike ended after 6 months, wages raised 50 percent 3 months later, bonus system revised
1924	Kauaʻi plantation	Filipino	Higher wages	13,000 strikers, ended after 8 months, 16 strikers and 4 police killed, 60 sent to prison, strike failed

Source: Takaki, *Pau Hana.*

occurred at Pioneer Mill Company and Oʻahu Sugar Company. The very first coordinated strike of workers on several plantations occurred on Oʻahu in 1909. Preceded by workers' formation of the Higher Wage Association, the "Great Strike" of 1909, as it was called, began at ʻAiea and spread to over 7,000 workers and all of the plantations on Oʻahu. The strikers received political and financial support from Japanese workers on the other islands, as well as help from some Japanese business organizations on Oʻahu. The workers demanded higher wages and the elimination of the system of differential wages based on ethnic group, which discriminated against the Japanese. In response, Japanese laborers were evicted from their plantation homes, and the owners brought in Hawaiian, Chinese, Portuguese, and Korean strikebreakers at twice normal wages. The strikers held out for four months before giving up and returning to work. But they had convinced the planters of the need to address their grievances, and three months after the end of the strike the HSPA abolished ethnic wage discrimination, effectively raising the pay of Japanese workers.

Before the Great Strike of 1909, all immigration was sponsored by the Kingdom, Republic, or Territorial government of Hawai'i. After the Great Strike, the HSPA took the lead in bringing workers to Hawai'i. In an effort to reduce the political power of Japanese workers, HSPA established offices in the Philippines and began to recruit more Filipino workers, including about 30,000 between 1906 and 1919, more than 70,000 during the 1920s, and almost 50,000 from 1930 to 1934. But most of these workers were single men, and many repatriated to the Philippines after they had earned enough to establish themselves in the homeland. Nevertheless, by 1931 there were 66,000 Filipinos in Hawai'i.[59]

In 1913 the total plantation labor force of 45,875 was composed of 24,282 Japanese, 8,101 Filipinos, 4,162 Portuguese, 2,174 Spanish, 1,524 Puerto Ricans, 1,402 Koreans, 1,126 Chinese, 1,040 Hawaiians, 663 Americans, 93 Russians, and 308 others.[60] By 1921 the islands' plantation labor force had declined to less than 38,000 as the result of increasing mechanization. In addition, its racial composition was changing, with fewer Japanese and more Filipinos. The classification of plantation laborers by ethnicity as of 1921 revealed that the majority of workers hailed from Japan and the Philippines; in all, there were 17,466 Japanese, 12,271 Filipinos, 2,500 Portuguese, 1,586 Chinese, 1,279 Puerto Ricans, 1,150 Koreans, 878 Americans, and fewer than 600 others.[61]

In the 1890s unskilled plantation labor was paid $12.50 to $15.00 per month (a month of work consisting of 26 ten-hour days). Women and children were used for hand weeding (hoe-hana), while men did heavier work like cutting, carrying, and loading cane by hand on cane-haul wagons.[62] But as a result of pressure from labor, especially the Great Strike of 1909, wages increased. By 1913–1914 all but the lowest job categories earned more than a dollar per day (see Table 3.5).

Labor accounted for about two-thirds the cost of field operations involved in producing cane in 1913–1914 (see Table 3.6). It makes sense, then, that

Table 3.5. Numbers of plantation workers and their average daily wages, 1913–1914

Occupation	Number	Daily wages
Cultivators (contractor)	6,885	$1.23
Field hands	10,899	$0.81
Cane cutters	3,173	$1.04
Cane loaders	2,045	$1.30
Teamsters	128	$1.15
Locomotive engineers	110	$2.03
Mill yard cane scale operators	81	$1.69
Mill chemists	21	$6.72
Mill chief engineers	6	$10.98
Sugar boilers	46	$6.29
Machinists	89	$2.64
Carpenters	216	$1.87
General mill laborers	2,492	$0.96

Source: U.S. Department of Commerce, *The Cane Sugar Industry,* 1917.

Table 3.6. Total cost per acre and labor cost for Hawaiian sugar plantations, 1913–1914

Operation	Cost per acre	Labor cost
Clearing	$1.85	78%
Steam plowing	$3.66	52%
Steam plow repairs	$0.96	45%
Mule plowing	$2.22	47%
Harrowing and furrowing	$2.09	58%
Preparing and ditching	$2.08	81%
Cutting seed	$2.25	87%
Hauling seed	$1.39	59%
Seed cane	$1.78	15%
Planting	$2.21	91%
Replanting	$0.61	86%
Total	$17.75	66%

Source: U.S. Department of Commerce, *The Cane Sugar Industry,* 1917.

plantations had every incentive to reduce labor costs by adopting labor-saving technologies. In addition, by purchasing labor-saving machinery, plantations helped field laborers to be more efficient. In 1912 the HSPA introduced a system of bonuses that allowed low-wage employees who worked at least 240 days per year to receive a share of plantation profits at the end of the season. For every dollar that the average price of sugar in New York exceeded $70 per ton, these workers received a 1 percent bonus based on their annual earnings. This amounted to a bonus of 13 percent in 1912, 1 percent in 1913 when prices were low, 5 percent in 1914, and 20 percent in 1915. During this period, about 15,000 laborers received these yearly bonuses.[63]

Cultivating and harvesting contracts were another tool used by plantations to increase labor productivity.

> It is the custom of plantation managers to enter into "cultivating contracts" with groups of their employees for the work necessary to produce the cane. . . . This system has proved mutually satisfactory to the companies and to the workmen. The workers have the incentive of greater earnings . . . and the company has the advantage of little or no supervisory expenses, knowing that the contractors will do their utmost in the work of irrigation, weeding, stripping, and fertilizing, as . . . the greater the number of tons of cane produced the greater their earnings.[64]

Contractors, who were typically Japanese, usually earned more than day laborers, though their income could be more variable. In addition, they could set their own hours and days of work, even working Sundays and bringing their wives and children with them.[65]

According to the U.S. Department of Commerce, for the fourteen cane crops produced between 1899 and 1912, one (unnamed) irrigated plantation found that contract labor produced 51.2 tons of cane per acre, compared with 43.3 tons of cane per acre produced by day labor. Plantations also used contractors for harvesting cane. A typical harvesting contractor had seven to 24 laborers in the field on an average day, with each laborer harvesting an average of six to eight tons of cane per day and earning $24–32 per month (of 26 workdays).

In 1914, the base wages for men engaged in field labor was typically $20 per month for Japanese and Chinese and $24 for Europeans. Because contracts with Philippine citizens required that they have a right to return passage to their homes at the end of their three-year contracts, their wages were typically $18 per month, though the Department of Commerce noted that "in practice, most of them break their contracts soon after they arrive, either by leaving the plantation altogether or by going to another plantation where they get $20 a month or more."[66] In addition, some plantations based pay rates on the number of days per month that the employee worked, increasing from $0.85 per day for men working less than 15 days per month to $1.00 per day for those working at least 24 days per month. In addition, employees received free housing, water, fuel, and (usually) medical services. Some plantations encouraged their employees to raise vegetables, pigs, and chickens.[67]

One complicating factor in analyzing plantation wages was that workers often shifted between jobs that paid different wages. "A man may be employed at the base wage of $20 to $24 a month hoeing out weeds for five days; then he may work with animals for a week at a higher wage; later he may spend a few days irrigating at a third rate; and for the rest of the month he may work in the mill or in mechanic's gang at a fourth rate."[68]

Because the plantations required large amounts of labor for both field and mill work, they developed communities, often near the mill but also among the cane fields, with plantation housing, stores, churches, schools, and recreation facilities. As different ethnic groups were recruited overseas and immigrated to Hawai'i, ethnic "camps" were established. By initially separating the ethnic groups, relatively homogeneous camp communities developed. Although the formation of groups along ethnic lines helped maintain ethnic identities, customs, and languages while assisting new immigrants, it tended to discourage cooperation among ethnic groups in labor disputes. Over time ethnic ties weakened due to intermarriage and socialization at work and school, gradually producing more ethnically diverse and integrated plantation communities.[69]

The industry recognized that it was imperative to maintain good living and working conditions for its labor force. Around 1905 companies began to "do away with undesirable and insanitary houses, and build homes for the married and unmarried in their stead." Led by the HSPA, the Industrial Service Bureau was formed in 1919 "for the promotion, primarily, of improvements in building and sanitation, as also such activities in amusements, recreation, and general welfare work . . . to improve labor and make for a contented people." Labor housing projects provided "villages of cottage homes . . . regular streets, alleys, sewage and water systems, playgrounds, community center, and all else that goes to make for a healthful settlement." Other services promoted by the bureau included free water supplies, free medical treatment with hospitals staffed by nurses at larger plantations, dairies to supply worker families with milk, day nurseries, home gardens, recreational areas, English language classes, programs to beautify community landscapes, and Girl and Boy Scout programs.[70]

In 1917 the U.S. Department of Commerce, in its comprehensive audit of the sugar industries in Hawai'i, Cuba, Puerto Rico, and Louisiana, reported:

> Several facts indicate that plantation laborers [in Hawai'i] receive more than a subsistence wage. The Japanese, including those not on plantations, send more than $1,000,000 annually to their home country through the post office, and are said to transmit a still larger sum through the Yokohama Specie Bank, which has a branch in Honolulu. The bank officials state that a large fraction of their remittances comes from plantation hands. Practically all Japanese in mercantile and small farming pursuits in the Territory (and their number is large and increasing) accumulated their original capital while working on a plantation. Spanish and Portuguese who arrive in Hawai'i from Europe penniless soon are able to migrate to California with money in their pockets, after paying the passages of themselves and their families to San Francisco. Not all plantation hands, by any means, are thus forehanded, and some of the thrifty and industrious are unable to accumulate because they are not physically able to stand the hard grind of plantation labor without frequent rests. . . . But many attractive little homes and thriving business undertakings have been established with the savings of plantation laborers, and their owners form the basis of a growing middle class.[71]

After World War I, wages did not keep pace with the cost of living. Despite the industry's strategy to import Filipino workers to counteract the Japanese influence, in 1920 these two ethnic groups cooperated to implement the first multiethnic strike. Like the Great Strike of 1909, their joint effort began with the formation of the Association of Higher Wage Question in 1917 to pressure the industry to raise wages. In 1919, largely ignored by the HSPA, Japanese workers formed labor organizations on each island, and these combined to form the Japanese Federation of Labor. During the same period, Filipino laborers organized the Filipino Federation of Labor. By the end of January 1920 both federations had gone on strike against the O'ahu plantations, taking with them 77 percent of the workforce—a total of 8,300 workers (see Table 3.4). Demands included an increase in wages from $0.77 to $1.25 per day for men and a minimum wage of $0.93 per day for women. Other demands were for an eight-hour day, an insurance fund to assist workers through illness and retirement, paid maternity leave, and an end to the policy that required workers to work a minimum of 20 days a month to receive bonuses based on the price of sugar. The workers complained that plantation work was so strenuous that only 40 percent of the workers could work 20 days per month, and they demanded that 75 percent of the bonus be paid monthly rather than all of it being paid at the end of the year. However, the HSPA refused to negotiate, attempting to pit the Japanese and Filipino federations against each other and paying strikebreakers $3 per day plus the bonus. The strikers held out for six months before returning to work. But three months later—in a response similar to the one that followed the Great Strike of 1909—the HSPA quietly increased wages by 50 percent, began paying bonuses on a monthly basis, and expanded social welfare and recreational programs.[72]

Shipping

Raw sugar produced in the islands was shipped primarily to San Francisco and New York. Between 1904 and 1914 the industry used a total of six routes:

- By sail or steam to San Francisco
- By sail or steam to San Francisco then overland to New York
- To New York by sail via Cape Horn
- To New York by steam via the Strait of Magellan
- To New York by steam and rail via the Isthmus of Tehuantepec
- To New York by steam via the Panama Canal (beginning in 1913–1914)

During most of this period the majority of the islands' sugar was shipped to New York. In 1904–1905 about a third of the sugar was carried in sailing ships, either to San Francisco or westward around Cape Horn to East Coast ports, but by 1914 all sugar was carried on steamships. The American-Hawaiian Steamship Company was founded in 1899 to carry sugar from Hawai'i to the United States and return with manufactured goods. Its steamships initially sailed around South America via the Straits of Magellan to reach East Coast ports. In 1907 the company began using the Isthmus of Tehuantepec route, which, with the exception of a short period when Mexico closed that route, carried the majority of the sugar to New York until 1914, when the Panama Canal opened for traffic. Throughout this period, the cost of shipping to San Francisco was $3–4 per ton. Shipping costs to New York varied from about $9.50 per ton via the Panama Canal, to $10–11 per ton via the Isthmus of Tehuantepec, to over $13 per ton overland by rail from San Francisco.[73]

Agricultural Research and Development

The Experiment Station of the HSPA had been formed in 1895 to help plantations improve their management and increase their yields and profits. Its annual report for 1898 described research on fertilization, irrigation, and 18 cane varieties. By 1905 it had added the Divisions of Agriculture and Chemistry, Entomology, and Pathology and Physiology, as well as six substations. The Experiment Station had 21 employees in 1909, including an entomologist, sugar technologist, pathologist, chemist, and agriculturist employees.[74]

Like most islands, the territory's agriculture was particularly susceptible to introduced pests, and the Experiment Station was a leader in introducing biological control, especially for weeds and insects detrimental to sugarcane and other agricultural crops. As of 1914 Experiment Station scientists were using biological agents to control scale insects, plant lice, mealybugs, and leaf-rollers that attacked sugarcane, coconut palms, and other trees. A number of insects were introduced to control lantana, an ornamental that had become a serious weed. Scientists also introduced egg parasites of sugarcane leaf hopper, a fly from New Guinea that preyed on and controlled sugarcane borer, and an egg parasite that controlled bean weevils on algarroba beans.[75]

A new Experiment Station headquarters building was completed in 1917, and during World War I the emphasis of the station's work turned temporarily to food production and processing in support of the war effort. In 1918 a Department of Botany and Forestry was established to house research on pathology, forestry, and pineapples, and by 1919 the Experiment Station had 32 employees. In 1925, its director attributed the record crop of 1925 to good weather, good management, an adequate labor supply, and the benefits of Experiment Station research, including (1) biological control of leaf hopper; (2) improved varieties like H 109, D 1135, and Yellow Tip; (3) control of mosaic disease through the use of healthy seed cane; (4) better viability of seed cane though careful selection; (5) control of field rats with chemical poisons; (6) more careful use of irrigation water; (7) more effective use of fertilizers; (8) better coordination of field practices; and (9) improved methods of factory operation and control.[76]

In 1926 Dr. A. J. Mangelsdorf, a prominent geneticist, was hired to lead the cane breeding program, and the following year a cane breeding facility was established on windward Oʻahu at Maunawili, where weather conditions caused cane to flower prolifically. In 1929 a cane quarantine station was established on Nāhiku, Molokaʻi, to protect the industry from diseases and insects that might enter with cane introductions.[77] An important advance was made in several sugar-growing countries in the early 1900s when it was discovered that hybrid varieties of sugarcane could be made by crossing within and between various *Saccharum* species. Up until this time pure *Saccharum officinarum* varieties were exclusively used by the planters in Hawaii and elsewhere around the world. The hybrids had the potential to provide disease and insect resistance and, along with advances in irrigation and fertilization, dramatically increase the yield potential of sugarcane. This is exactly what happened with the selection of the hybrid H 109 in 1905, which eventually replaced Lahaina, the most prominent variety used by Hawaiian planters. H 109 was planted on more than 100,000 acres by the 1930s.

In 1901, three years after annexation, Congress established the federally funded Hawaii Agricultural Experiment Station (HAES) with headquarters on Oʻahu. Leaving sugarcane research to the Experiment Station of the HSPA, HAES focused on agricultural diversification.[78] Early tobacco research demonstrated that "with attention to varieties, soils, curing, and fermentation a product was secured that was given high rank by experts." New forages were successfully grown, and more shipping-tolerant banana varieties from Central America were introduced. Early citrus research focused on replacing imports from California with local production. By 1913 E. V. Wilcox, Special Agent in Charge of the Hawaii Agricultural Experiment Station, was promoting farmer cooperatives to reduce costs to small farmers and foster better agricultural practices.[79]

The value of the forests in slowing runoff and controlling erosion had become evident by the turn of the twentieth century. In 1902 the Irish civil engineer M. M. O'Shaughnessy, writing about the importance of irrigation to the sugar industry, indicated that "the rugged character of the mountain chains, as well as forest growth, have important bearing on precipitation . . . [and] the folly of wasteful forest destruction is now pretty well understood."[80] As a result, in 1903 the Territory of Hawaiʻi, with the strong backing of the HSPA, established a Board of Agriculture and Forestry, predating the USDA Forest Service by one year. In 1904 the first Territorial Forester was hired, and forest reserves were created to protect upper watershed areas. These reserves were managed by fencing out domestic livestock and wild game, eliminating feral hogs, and planting native and exotic tree species.[81]

Overall, the first and second decades of the 1900s were a time of considerable learning and consolidation of knowledge in Hawaiian agriculture. University life on the islands began with the founding of the College of Agriculture and Mechanic Arts in 1907; its name was changed to the College of Hawai'i in 1912 and then to the University of Hawai'i in 1920. With the establishment of a full-fledged university came the inauguration of a state-funded extension service; additional funding from the federal government allowed the Hawai'i Cooperative Agricultural Extension Service to launch in 1928.[82]

Diversification of Hawai'i's agriculture was a natural next step for the islands following annexation. Recognizing the potential for small-farm coffee production, Japanese coffee farmers established the Kona Japanese Coffee Producers Association in 1898 to improve processing and marketing. Within a dozen years Japanese coffee farmers made up 80 percent of the total farming population in Kona,[83] and by 1912 the territorial government had developed a policy to sell farm land on "exceedingly easy terms" in an attempt to create a class of smaller farmers. Land was sold at auction, with normal prices of $2–5 for pasture land, $5–25 for land that could be used for cultivation of vegetables, and $28–60 for sugarcane land. These prices were about 25 percent of the estimated market values, with easy payment terms of 5 percent down and 5 percent per year for 20 years, at zero interest, though purchasers had to agree to develop the land.[84] This program led to predictions that "the future agriculture of Hawai'i will largely be made up of small farms and diversified agriculture. These farms will be owned by the man who tills the soil—the man who makes his home on the land."[85] Boosters of diversified agriculture pointed out that large amounts of beef, pork, butter, eggs, tobacco, vegetables, horses, mules, and feed grains were being imported but could be produced profitably in the Islands. They also noted that the small-scale sugarcane farmers in Queensland who sold their cane to large mills could be good models for farmers in Hawai'i.[86]

Agricultural diversification with pineapple eventually proved successful, initially with small holders but eventually with large plantations. Pineapple is said to have been introduced into the Kingdom in 1813 but was not developed commercially until the 1880s. John Ackerman and Waldemar Muller canned pineapple commercially in Kona in 1882. Captain John Kidwell, a pioneer of the industry, began experimenting with pineapple in 1885, planted the Smooth Cayenne pineapple variety near Pearl Harbor in 1890, and two years later joined with John Emmeluth to build a pineapple cannery in Waipahu. In 1897, 150,000 pounds of pineapple worth $14,000 were exported from the Kingdom. Alfred Eames started selling fresh pineapple in 1900; his early success led to the eventual founding of Del Monte Fresh Produce (Hawai'i), Inc. In 1901 James Dole founded the Hawaiian Pineapple

Company, which grew into a much larger venture within just a few years. In 1905 the company packed 125,000 cases of pineapple; a year later it built the Iwilei Cannery. An improved pineapple processing machine was patented by Dole employee Henry Ginaca in 1914, further stimulating production of canned fruit. Around the same time, the Pineapple Packers Association (PPA) established an alliance with the HSPA to conduct pineapple breeding research, and in 1923 the PPA formed the Pineapple Research Institute.[87]

In 1914 the total value of agricultural exports from the islands was just over $41.5 million, of which raw sugar contributed $33 million and pineapples $5 million. Other crops lagged far behind, including coffee ($0.83 million), rice ($0.18 million), and bananas ($0.12 million). A number of problems contributed to low production and exports of crops other than sugar and pineapples. The Board of Agriculture and Forestry lamented the "lack of suitable rice land and the absence of Chinese labor" required to produce rice. The pink bollworm had caused cotton to be "practically abandoned," and lack of financing, difficulty in curing, and fires that burned drying bars had prevented the tobacco industry from developing in Hawai'i. Coffee growing had "existed for a number of years as a minor industry," but it was producing only "very moderate returns." Finally, bananas produced in the islands were "poor shippers" and could not compete with those grown in the West Indies and Central America, except perhaps in the San Francisco market.[88]

Figures provided by the U.S. Department of Agriculture showed that, a decade later, in 1924, the value of sugar and pineapple exports to the United States had increased substantially, while other crops continued to lag: raw sugar ($73.9 million), canned pineapple ($28.2 million), coffee ($0.4 million), and bananas ($0.2 million).[89]

Maui Sugar

Industry optimism resulting from annexation in 1898 led to some aggressive business practices on Maui. For example, the first railroad, the Kahului and Wailuku Railroad Company, had been formed in 1879 and became Kahului Railroad Company (KRR) in 1881 with operations linking Wailuku, Kahului, and Pā'ia. However, in 1897 Claus Spreckels, foreseeing the upcoming expansion of the sugar industry, organized Maui Railroad & Steamship Company (MR&SC) to compete with KRR. By transferring HC&S

land to MR&SC, he was able to cut off access by KRR to the waterfront. Meanwhile, Alexander & Baldwin, in cooperation with James B. Castle, purchased control of HC&S from Spreckels, and Baldwin became its president. To secure its access to the port, HC&S purchased both KRR and MR&SC in 1899 and merged the two railroads into KRR. In this way, Alexander & Baldwin gained control of all rail and port facilities on Maui.[90]

In 1898 Pā'ia Plantation installed a new 9-roller mill, engines, and pumps. Soon thereafter, Alexander & Baldwin (A&B) was serving as the agent for Pā'ia Plantation Company, 'Haikū Sugar Company, HC&S, and KRR. In 1900 A&B was converted from a partnership to a corporation serving as agent for major plantations, and the corporation's holdings included shares in California and Hawaiian Sugar Refining Company, 'Ōla'a Sugar Company on the island of Hawai'i, Hawaiian Fertilizer Company, Union Feed Company, and Nāhiku Sugar Company on Maui.[91] Around the same time— and with the optimism characteristic of the era—Henry Baldwin and his partner Lorrin Thurston formed Kīhei Sugar Company on some dry ranch land they owned in south-central East Maui. The plan was to raise cane to be milled at the HC&S mill at Pu'unēnē. Kīhei Plantation would depend on purchased ditch water and its own wells for irrigation. By 1903 the plantation was receiving "a good supply of water" from both the HC&S and Pā'ia and 'Haikū (Spreckels) ditches, and anticipated paying Maui Agricultural Co. $5 per million gallons for water in 1904.[92]

In the late summer of 1901 the *Maui News* reported that "the pride of Maui and the wonder of the world, [HC&S's] new 500 ton Pu'unēnē Mill, is nearing completion." The mill began grinding in 1902, and, according to an article in the *Maui News* dated January 24, 1903, the Pu'unēnē mill had "turned out 250 tons of sugar per day, every day this week, which is the world's record to date." The mill, claiming to be "the largest sugar mill on earth," had all its milling operations under one roof, recycled the mill's process water for irrigation, and burned its bagasse to fuel its boilers. Five to six hundred narrow gauge traincars hauled cane to the mill each day, and over three cars of raw sugar in sacks were taken to the wharf each day for shipment to California for refining.[93]

In response to the landholding and plantation size limits put in place by the Organic Act, in 1903 Maui Agricultural Company was formed by combining two well-established plantations, 'Haikū Sugar Company (formed in 1858) and Pā'ia Plantation Company (established in 1883). A year later, Kīhei Sugar Company was divided into five smaller plantations, each less than 1,000 acres in accordance with the Organic Act: Kula Plantation Company, Makawao Plantation Company, Pūlehu Plantation Company, Kailua Plantation, and Kalianui Plantation Company. Cane from the seven entities was initially ground at the Hāmākuapoko and Pāʻia factories, but in 1906 the equipment from the two factories was combined at Pāʻia, creating a tandem mill. This was converted a few years later to a 21-roller mill. The seven small plantations operated separately until 1921, when the Organic Act was repealed and they were all merged into Maui Agricultural Company.[94]

The average sugar yield for the entire territory in 1910 was 4.69 tons of sugar per acre, with average yields of 6.27 and 3.06 tons sugar per acre for irrigated and unirrigated lands, respectively. As an industry average, 7.99 tons of cane was needed to produce one ton of sugar. Maui, because of its well-developed irrigation systems and high solar radiation levels, was outpacing the other islands in sugar production per acre, yielding 7.39 tons sugar per acre in 1910, compared with 6.26, 4.90, and 3.03 tons sugar per acre on Oʻahu, Kauaʻi, and Hawaiʻi, respectively.[95]

Maui's plantations differed substantially in size, and their production varied from year to year, due primarily to rainfall and the amount of water available for irrigation. In 1912, a good year, and 1913, a dry year, the four largest Maui plantations, all benefitting from well-developed irrigation systems, produced net profits per ton of sugar that were, in general, substantially better than the average for island plantations (Table 3.7).[96]

Maui's sugar companies were forced to adapt to shortages of food and fuel during World War I. ʻHaikū Plantation grew corn and milled it for the public. In 1917 the Maui Agricultural Company began making ethanol from molasses, using it as fuel for its fleet of cars and trucks and continued to do so until 1923, when falling gasoline prices made it uneconomical.[97] In addition, in 1918 the HC&S Central Power Plant was completed, serving all of Central Maui until Maui Electric Company constructed its plant at Kahului after World War I. Also in 1918, HC&S purchased $3,245 worth of tractors, truck, and automobiles, beginning the process of gradually reducing its dependence on horses, mules, and steam power for transportation and plowing. And the investments didn't stop there. Responding to good profits due to high sugar prices following the war, HC&S built a theater, a Filipino clubhouse, and a new plantation dairy. It also began extending electrical service to all the plantation villages. Harry Baldwin, manager of Maui Agricultural Company, was elected the Territory's delegate to the U. S. Con-

Table 3.7. Sugar yields (short tons) and net profits for Maui plantations in 1912 and 1913

Plantation	1912 Yields tons/year	1912 Profits $/ton	1913 Yields tons/year	1913 Profits $/ton
Hawaiian Commercial & Sugar Co.	60,010	$36.23	50,310	$17.56
Maui Agricultural Co.*	34,612	$33.45	24,633	$6.45
Pioneer Mill Co.	28,335	$25.51	27,804	$13.54
Wailuku Sugar Co.	16,775	$33.09	13,988	$13.13
Kaʻelekū Plantation Co.	4,949	—	4,938	—
Kīpahulu Sugar Co.	2,197	—	1,408	—
Olowalu Co.	1,707	—	1,738	—
43 sugar companies in islands	548,351	$21.74	507,162	$7.00

Source: Thrum, *Almanac,* 1914.
* Includes ʻHaikū Sugar Company, Pāʻia Plantation Company, Kula Plantation Company, Makawao Plantation Company, Pūlehu Plantation Company, Kailua Plantation, and Kalianui Plantation Company, which were merged into Maui Agricultural Company in 1921.

gress in 1922. The following year a severe tidal wave caused extensive damage in Kahului, but progress continued in 1924 with the completion of the Kaheka hydroelectric plant.[98]

After ʻHaikū Sugar Company, Pāʻia Plantation Company, Kula Plantation Company, Makawao Plantation Company, Pūlehu Plantation Company, Kailua Plantation, and Kalianui Plantation Company were absorbed into Maui Agricultural Company in 1921, Maui was home to seven plantations. By 1925, these plantations had 40,000 acres planted to cane. Most fields at that time grew two-year cane; therefore, in the 1925 season, about 20,000 acres of cane were harvested. The plantations harvested for an average of 141 days, and the fields yielded an average of 63 tons of cane per acre, with about 13.5 percent of the cane extracted as raw sugar. The 170,000 tons of sugar and 1,258,000 tons cane produced on Maui made up 22 percent of the sugar and 20 percent of the cane produced in the territory that year.[99]

Kīpahulu Sugar Company shut down following the harvest of its 1925 crop. The remaining Maui plantations were Pioneer Mill on the western coast of West Maui, Olowalu Company just south of Pioneer Mill, Wailuku Sugar Company on the eastern slopes of West Maui, Hawaiʻi Commercial and Sugar Company and Maui Agricultural Company on the isthmus and western slopes of East Maui, and Kaʻelekū Plantation Company on the eastern coast of East Maui.[100]

In the 1920s Hawai'i sugar plantations were substantial, largely self-sufficient communities composed of managers, skilled and unskilled laborers, and their families, almost all of whom lived on the plantations. Each plantation supported its community of workers by providing housing, medical and social services, and much of the food and fuel needed by its workers. This required the companies to manage a variety of agricultural production and processing activities. For example, HC&S had 3,000 employees and a total population, including children, of 7,000. It owned and maintained 26 villages (or camps) scattered across the plantation to house its workers, and it operated 12 nurseries where mothers could leave their children while at work. It also maintained recreational amenities such as a clubhouse, tennis court, bowling alley, and swimming pool; Korean, Hawaiian, Japanese, and Catholic churches; and a hospital with two physicians and four nurses. To feed its workers and their families the plantation slaughtered 600 head of cattle, primarily Angus and Durham, and hundreds of hogs and sheep. It had its own dairy and grew garden vegetables. Cattle were fattened with a mixture of plantation-grown alfalfa, corn, molasses and other feeds, including cane tops, sweet potatoes, and roasted algaroba beans. "The company dairy distribute[d] an average of 1,000 quarts of bottled, sterilized milk daily to its employees at less than cost" from its 66 Holstein and Jersey cows. Duroc hogs bred to Giltner boars provided pork, while excess milk, buttermilk, and offal from the slaughterhouse were fed to the hogs. The company's 1,000 head of horses and mules were cared for in seven stables distributed across the plantation. It bred its own saddle horses but purchased work teams from Parker Ranch.[101]

In 1926 HC&S had 85 miles of railways to move its cane to the factory. Like other plantations, HC&S continued to plant improved cane varieties, with Lahaina its predominant variety from 1913 to 1920. But increasing amounts of H 1135 were grown in the early 1920s, and H 109—the first variety resulting from the HSPA breeding program—became the predominant variety from 1923 through 1928.[102]

Water Development

Development of irrigation ditch systems began in earnest on the major islands in the last half of the nineteenth century, with Kaua'i plantations leading the way. Among the first were Lihui Plantation's Rice Ditch in 1856 and Hanamā'ulu Ditch in about 1870, several small ditches at Grove Farm in the 1860s and Kōloa Sugar Company's ditch (also known as Dole's "water lead") in 1869. But Maui plantations soon followed, with Alexander & Baldwin's (Old) Hāmākua Ditch in 1878 and Spreckels's (Old) 'Haikū Ditch in 1879.[103] Construction of irrigation systems accelerated in the 1880s and 1890s as plantations reinvested profits from tariff-free sugar. By 1898, 55,973 acres of sugarcane were irrigated—slightly more than half of the islands' sugar lands—and by 1902 numerous ditches had been built on Maui, O'ahu, and Kaua'i.[104]

In the early 1900s the Hawaiian sugar industry developed what soon became known as the "Maui-type well" to "skim" the water from the freshwater lens of the basal aquifers under each island while minimizing contamination from the more salty transition zone below. The first of these Maui-type wells was installed at Kīhei Plantation on the dry leeward side of East Maui in 1900. These wells consisted of a vertical shaft dug from the land surface to the top of the basal aquifer. A large pump installed just above the top of the aquifer drew water from a sump extending downward into the freshwater lens. One or more tunnels were then dug horizontally from the sump for varying distances to allow the water in the saturated layers of lava to infiltrate into the tunnel in response to pumping.[105]

The length of lateral tunnels required by Maui-type wells depended on the permeability of the lava layers comprising the upper part of the freshwater lens. For example, on East Maui the Honomanū basalts are more permeable than the Kula lavas; therefore, Maui-type wells in the Honomanū typically had higher yields than those in the Kula.[106] The original pumps for Maui-type wells were driven by coal-powered steam plants located on the surface and powering steam turbines attached to the pumps at the bottom of the vertical shaft. But these steam plants were soon replaced by electric motors coupled directly to the pumps. The early wells were located near the coast, where the depth to the water table was lowest; however, the freshwater lens was also thinnest near the coast, and as motors and pumps became more powerful, plantations dug new wells farther inland where the freshwater lens was thicker and the risk of drawing water from the brackish transition zone was reduced.[107] The danger of saltwater intrusion into irrigation wells due to excessive pumping was well appreciated by the beginning of the twentieth century. O'Shaughnessy cautioned that "it is a great mistake . . . to concentrate the pumping stations and draw too much water from a small area, as there is a liability to . . . draw sea water into the wells if the surface of the fresh water is lowered too much."[108]

Irrigation was expensive. For example, in 1903 Kīhei Plantation Company spent over $26,000 to buy ditch water, almost $47,000 in labor to irrigate the crop, over $55,000 to run its pumps, and another $24,000 to repair them. In comparison, the plantation spent only $46,000 for fertilizing, $4,000 for steam plowing, $7,500 for horse and mule expenses, and $18,000 for cutting, loading, and hauling cane. Though pump water was often needed to supplement low ditch flows, the minerals and sand it often contained caused problems. Kīhei Plantation was forced to repair two Worthington pumps—responsible for delivering 6 mgd—whose "plungers and bushings [had] been so badly cut and worn . . . owing to the amount of grit and sand passing through the water cylinders . . . when the water level was low." Good quality water was also important to supply the plantation's steam engines. In 1903 the plantation began using ditch water rather than the more mineralized well water to feed the boilers of its steam-powered pumping plants, steam plows, and locomotives.[109]

The sugarcane industry boasted 1,500 miles of irrigation canals (70 miles of which were tunnels) and 250 reservoirs by 1912.[110] A total of 428 wells were also in use, and "all the water from subterranean sources [was] pumped up by steam pumps of a total of 27,000 horse-power to a height of some 500 ft. [maximum], and [flowed] through 70 miles of iron pipes 16 to 54 ins. in diameter."[111]

Of course, development of irrigation systems that collected and transported large amounts of surface water required legally secure water rights. By 1933 Wadsworth could write that

a large part of modern [sugarcane irrigation] development has been made by securing such rights by purchase of the ahupuaʻa and the subsequent transfer of available water to areas which in early days had never been irrigated. Difficulties, of course, arise in the determination of the amount available for diversion in view of the ancient and well-established rights below the point of diversion. The same indefiniteness exists when old kuleana rights are purchased. In general the amount of water involved in such cases has been determined upon testimony of local witnesses, although the actual measurement of the water required for satisfactory taro production under local conditions is now a more common method. Waters arising upon government lands are leased for definite periods upon the terms secured at a formal auction."[112]

Even before the turn of the twentieth century, plantations were being warned that they were wasting valuable irrigation water. O'Shaughnessy estimated that only about a third of the irrigation water applied reached the roots of the sugarcane crop—"where it would do the most good"—because of "leaky reservoirs, ditches, and unequal distribution" in the fields. He also complained that "no accurate measurements have been made showing the loss from seepage in these ditches, which is to be regretted, but rough measurement, made last year showed a forty percent loss in the Hāmākua ditch" and in several other ditch sections between four and 12 miles long.[113]

O'Shaughnessy further observed that "the loss of gravity [mountain ditch] water is bad enough, but the loss of water pumped to elevation at a heavy expense for fuel and pumping machinery is so serious a matter that it should be stopped." He recommended investing in more advanced and leak-resistant ditches or flumes rather than incurring "this constant loss" for fuel and pump depreciation and maintenance. To illustrate his point, he noted that "a three foot diameter semi-circular galvanized iron flume laid on a grade of ten feet per mile delivers without loss 10,000,000 gallons daily on the Pioneer Plantation on Maui on porous ground."[114]

O'Shaughnessy also recommended installing "well constructed weirs with beveled edges" to measure ditch flows. He provided the industry with tables to calculate daily ditch flows from the depth of water in weirs, and he noted that "measuring instruments can be purchased from instrument makers such as Leitz or Sala of San Francisco which are self-recording, and sheets can be taken off weekly which will show the depth of water at each hour during that period." Finally, he stressed "the necessity of building good reservoirs wherever tight and economic sites exist," and he stated that each pump should be connected with a "distributing reservoir of at least a size to hold a day's run of the pump." The reservoirs "should be tight" like new ones at Pioneer Mill Company that were "paved with stone and grouted with cement." He recommended that "all earthen dams should be built up in layers one foot thick, thoroughly tamped, wetted, and rolled. The inner slopes should be flat at least 2½ to 1 and paved with stone, and ample stone-paved wasteways should be made through the original formation away from the dam to allow freshets to escape." He opposed building dams over 50 feet high because "the subsoil and formation are so porous and uncertain."[115]

By the 1920s plantations had taken to heart the need to reduce leakage to save water. In addition, the practice of lining ditches substantially reduced ditch erosion and maintenance costs, including removal of weeds that could grow in the ditch and obstruct water flows. Lining also reduced the roughness of the ditch walls and bottom, allowing water to flow more rapidly and increasing the volume the ditch could carry. For example, J. H. Foss,[116] in a review of ditch lining in the Hawaiian sugar industry, noted that "the maximum velocity [of water flow] for unlined ditches should be about three feet per second, preferably not greater than two feet," to avoid excessive erosion of the ditch. While "masonry-lined ditches withstand velocities of from six to ten feet per second," he warned that the water in masonry-lined ditches should be "comparatively free from grit" since "grit-laden water running at a velocity of about four or five feet a second eroded the floor of the ditch about one inch in eight years' time."

A number of different types of lining were used. Cut stone was frequently used in rough country, where it was difficult to bring in large amounts of concrete. Plaster lining about three-fourths of an inch thick was used extensively between 1912 and 1917, but it was considered "more or less of a failure" because "after cracking the water gets under the edge, and a high velocity will cause the floor lining to roll up in a manner similar to a roll of carpet, and be carried down the ditch." Concrete-lined ditches, especially those cast in place with crushed rock rather than coral sand, were the most durable.[117]

Maui's Water Resources

By the late 1800s sugar plantations on Maui and elsewhere in Hawai'i clearly understood the importance of irrigation for the drier, leeward sides of islands. For many years emphasis was placed on developing irrigation sources and reducing the labor costs of managing furrow irrigation. Plantations were less concerned about the total amount of irrigation water applied per crop or the amount of cane or sugar produced per unit of water applied. Since plantations understood very little about the amount of water required by the crop, they often applied excessive amounts. In 1889 crops at Spreckelsville and Hāmākuapoko were estimated to have received between 1.01 to 1.23 cm/day. From 1912 to 1916, Wailuku fields received an average of 0.89 cm/day. HC&S applied even more—an average of 1.22 cm/day. But by 1958 crops at Mā'alaea, Pūlehu, and Kāheka were being irrigated at rates of 0.56 to 0.58 cm/day, a reduction of about 50 percent from the late 1880s.[118]

Soon after gaining control of HC&S in 1898, Alexander and Baldwin had their Hāmākua Ditch Company (formed in 1876) begin digging the Lowrie Ditch, named for William Lowrie, manager of HC&S's Plantation and mill at Spreckelsville. Dug by Japanese laborers in just over a year (1899–1900) for a cost of $271,141, the Lowrie Ditch was 22 miles long and consisted of 75 percent open ditch, 74 tunnels, 19 flumes, and 12 siphons. It had a capacity of 60 mgd and was designed to irrigate 6,000 acres. The Lowrie Ditch crossed land of both the Pāʻia Sugar Company and the ʻHaikū Sugar Company, with Pāʻia receiving 5.5 percent and ʻHaikū 4.5 percent of the ditch's water as compensation.[119] The Lowrie Ditch was located above (*mauka*) the old Spreckels (ʻHaikū) Ditch and below (*makai*) the Hāmākua Ditch. It took water from two sources, a reservoir near Pāpaʻaʻea Stream (that was fed by two ditches) and Kailua Stream (that fed the Spreckels Ditch).[120] In 1902 O'Shaughnessy observed, "In the last two years this [Spreckels] ditch was intercepted at Kailua by a new [Lowrie] ditch run on a grade of four feet per mile, twenty-two miles long, which delivers water at 450 feet elevation, thus obviating the expense of pumping to the higher level."[121]

Around this time Maui plantations were also developing their groundwater resources. The *Maui News* reported in the fall of 1901 that "two new pumps [would] be erected, one for Spreckelsville and the other for Pāʻia Plantation. The pumps will have a capacity of ten million gallons and will be in operation in time for next year's crops."[122]

O'Shaughnessy's studies of the ditches and wells on Maui landed him the job of designing the next large Hāmākua Ditch Company project, the Koʻolau Ditch. This project was built in 1904–1905 and extended the company's system of ditches from the eastern end of the New Hāmākua Ditch 10 miles eastward across and through extremely rugged terrain. Built at a cost of $511,330, it had 7.5 miles of tunnels, 2.5 miles of open ditch, and a capacity of 85 mgd. Its 38 tunnels were 8 feet wide and 7 feet tall and were built by Japanese laborers in 18 months working three 8-hour shifts a day. In order to double the speed of the work, tunnels were simultaneously dug from opposite sides of barriers, requiring precise surveying to assure that they met underground. Six inches of concrete was used to line 4.5 miles of the tunnels. The project also required construction of 4.5 miles of wagon road and 18 miles of trails for pack animals to bring supplies and workers to construction sites along the ditch.[123]

In 1908 Alexander and Baldwin formed the East Maui Irrigation Company (EMI) from the Hāmākua Ditch Company and gave it responsibility of developing and managing the surface water supply system from Māliko Gorge (where HC&S Plantation began) eastward to Nāhiku. The EMI proceeded to spend over $385,000 lining and improving the Koʻolau Ditch. By 1913 the EMI system consisted of the Koʻolau Ditch on the eastern end of the system. At Alo Stream near its western end, the Koʻolau Ditch joined the New Hāmākua Ditch, which continued westward toward the west end of the EMI system at Māliko Gorge. In addition, Koʻolau ditch water could be diverted to Pāpaʻaʻea stream and reservoir, just west of Alo, where it could flow into the Lowrie Ditch or the ʻHaikū (Spreckels) Ditch. The highest elevation ditch was Kaluanui Ditch, which crossed Māliko Gorge and contributed its waters to the New Hāmākua Ditch. From mauka to makai (high to low elevation), the five EMI ditches were the Kaluanui, New Hāmākua, Old Hāmākua, Lowrie, and ʻHaikū ditches. The EMI ditches continued across Māliko Gorge onto HC&S Plantation land, where their flows were managed by the plantation to irrigate its fields. In addition, in 1913 the HC&S irrigation system included wells, one near Puʻunēnē and two near Kīhei.[124]

EMI continued expanding its system, producing ditches with ever-larger capacities, the first being the 100-mgd New ʻHaikū Ditch completed in 1914. Mostly tunnel, it was partially lined and more than 54,000 feet long. The Kauhikoa Ditch, almost 30,000 feet long with a capacity of 110 mgd, was completed in 1915. Finally, the Wailoa Ditch, over 51,000 feet long with an initial capacity of 160 mgd, was completed in 1923. EMI then began digging water development tunnels that penetrated into water-bearing rocks and directed the water they captured to the ditches.[125]

East Maui Irrigation's ditch system soon included 388 separate intakes diverting water from streams and other sources, 50 miles of tunnels, 24 miles of ditches, 12 inverted siphons (that carried water down into gorges and back up the other side, where they emptied into canals or tunnels), numerous flumes that carried water straight across smaller gorges, several small reservoirs and dams, and 62 miles of private roads to access the system. EMI owned 18,000 acres but collected water from a total watershed of 56,000 acres, the remainder owned by the Territory of Hawaiʻi, which granted EMI four licenses to harvest and transfer the water. The price EMI paid for the water depended on the price of sugar and the fraction of the total rainfall in the watershed that fell on territory lands.[126]

In a retrospective chronicling more than two decades of irrigation advancements, a contributor for the *Honolulu Advertiser* wrote that HC&S had "a remarkable series of water developments . . . not the least of these was the bringing of Nāhiku waters to Puʻunēnē and Pāʻia in the great Wailoa ditch—a project which has been 20 years in the making and was only completed last May [1923]." The Wailoa, meaning "the water a long way off," was composed of "30-odd miles of underground tunnels, concrete-lined and arched for the entire distance."[127]

The total water requirements of the HC&S Plantation in 1923 were reported by the *Advertiser* to be over 140 mgd, of which 100 million gallons could be supplied from several wells dug to the basal aquifer under East Maui. Varying amounts of surface water came from the old Spreckels Ditch, the Lowrie Ditch, and the new Wailoa Ditch. Three major pump stations were in use. Pump Station 2 was powered by two generators with a combined power of 3.5 MW driven by an oil-fired steam turbine. Pump Station 1 in Kīhei had a capacity of 12 mgd; half the water was pumped with electric power and half with a combination of diesel and Sulzer cycle engines. The underground chamber of Pump Station 1 had recently been remodeled to provide better ventilation, and an elevator capable of holding eight to 12 men had replaced the old "bucket" previously used to lower and raise personnel. Pump Station 5 had two Riedler steam pumps rated at 10–12.5 mgd that had been operating for 20 years.[128] Well 7, developed in 1926, was—at 125 feet deep—one of the most productive water wells in the world.[129] This allowed the company to more than triple the acreage of sugarcane it harvested between 1900 and 1930 (see Table 3.8).

In the 1920s the EMI supplied mountain water from its ditches to both HC&S and Maui Agricultural Company (MAC). Since the flows depended on mountain rainfall on windward slopes of East Maui, they could decrease substantially in years when the trade winds were weak. Flows in 1926 and 1927 illustrate how weather conditions could affect HC&S and MAC. For example, Wailoa Ditch, the highest ditch at 1,200 feet, supplied 53 percent of its water to HC&S and 47 percent to MAC. In 1926, an exceptionally dry year, it delivered only 27.8 billion gallons, but its flows recovered to 43.7 billion gallons in 1927. The Kuhikoa (Hāmākua) Ditch, at an elevation of 1,000 feet, delivered all its water to MAC (2 billion gallons and 8.2 billion gallons in 1926 and 1927, respectively). The low elevation ʻHaikū Ditch (comprising the Lowrie, Central, and Manuel Luis ditches), at an

Table 3.8. Hawaiian & Commercial Sugar Company harvested acreage and sugar yields, measured every five years between 1895 and 1930

Year	Acres harvested	Tons sugar/acre
1895	2,570	2.64
1900	2,484	7.19
1905	4,827	8.16
1910	6,488	8.76
1915	6,572	8.64
1920	6,547	8.72
1925	6,715	10.09
1930	7,605	9.53

Source: Gilmore, *Sugar Manual,* 1931.

elevation of 700 feet, provided all its water to HC&S, delivering 5.9 billion gallons in 1926 and 15.1 billion gallons in 1927.[130]

In an article from the winter of 1927 the *Honolulu Advertiser* reported that although 1926 had been the driest year on record for HC&S, the plantation managed to post a profit of $1.33 million, or 13.3 percent on capital stock valued at $10 million. This profit was generated by 63,555 tons of sugar harvested from 6,889 acres with an average yield of 9.28 tons of sugar per acre. "During the year the sum of $354,441.68 was expended on improvements, the major portion of this amount being spent on pumping stations, ditches, and power plant with the intent of placing all but 10 percent of the cane area where it can be reached by pumps in dry years." The use of well water by HC&S varied from 15 billion gallons in 1923 to 32 billion gallons in the drought year of 1926. Total irrigation water supplied to HC&S cane crops ranged from 56 billion gallons in 1923 to 72 billion gallons in 1927. The cost of pump water was about $5 per million gallons, and EMI ditch water cost the plantation $7–11 per million gallons.[131]

Small reservoirs were used to help distribute ditch and well water, providing overnight and temporary storage needed to manage the complicated delivery of water to cane fields all across the HC&S and MAC plantations. EMI had seven reservoirs with 200 million gallons capacity, MAC had 26 reservoirs capable of holding 340 million gallons, and HC&S had 20 reservoirs with total storage of over 500 million gallons. Many of the reservoirs were constructed on permeable soils and were prone to leak substantial amounts of water. Seepage rates were a huge concern for Hawai'i's sugar growers, so in 1927 F. E. House conducted seepage studies on the HC&S reservoir at Puʻunēnē, measuring flow through 1.0 square foot of soil 1.0 foot thick with a hydrostatic head of 8.5 feet. Depending on the location in the reservoir from which the soil was taken, initial seepage rates ranged from 0.34 to a whopping 254 gallons per square foot per day. But treatment of the soils with 500 pounds of salt per 1,000 square feet reduced these seepage rates to 0.05–0.78 gallons per square foot per day, a 300- to 600-fold reduction in some of the worst cases.[132]

Seepage from ditches could also be a serious problem on Maui's pervious soils, and tunnels through permeable basalts and lavas could also leak. As a result, EMI, HC&S, and MAC all had extensive ditch-lining programs. By 1928 EMI had over 50 miles of concrete-lined ditches and tunnels. HC&S had 73,000 feet of ditch lined with 3 inches of concrete; 23,000 feet of ditch lined with concrete plaster reinforced with chicken wire; and 25,000 feet of redwood flumes 24 inches wide and 11 inches deep. MAC had 16,000 feet of lined ditch and was experimenting with different types of redwood flumes to distribute water from its canals to areas within its fields, with the hope of reducing irrigation labor costs while maintaining or improving cane yields.[133]

Not only were collecting and transferring water important, so was gaining information on how much water was available throughout the ditches and reservoirs of the EMI and plantations. To this end, EMI, the U.S. Geological Survey, and HC&S had 60, 40, and 16 ditch control stations, respectively, to record water levels and estimate flows.[134]

West Maui

During the early decades of the twentieth century the sugarcane industry on West Maui was also developing its surface and groundwater resources. The *Maui News* reported in 1901, "Many streams have been diverted and reservoirs made during the last five years on the Wailuku and Pioneer Plantations on Maui." In addition, California miners were making rapid progress on the HC&S tunnel in ʻĪao Valley on the windward side of West Maui.[135]

Work on the Waiheʻe Canal, built jointly by Wailuku Sugar Company and HC&S, began in 1905 and took 18 months to complete. Up to 50 mgd flowed from Waiheʻe Stream above the older Waiheʻe Ditch. With a length of 10.6 miles, it cost $160,000 to construct and had 22 tunnels that totaled 16,539 feet in length; 39 flumes totaling 2,764 feet; 35,549 feet of

Table 3.9. Ditches serving Pioneer Mill Company

Name	Date	Ave. Flow	Capacity	Comments
		Millions of gallons/day		
Honokōwai (original)	1898		10	semicircular iron flume
Honokōhau	1904	20	35	owner, Honolua Ranch
Honolua (Honokōhau)	1913	18–30	50–70	replaced 1904 ditch, relined 1923–1928
Honokōwai (rebuilt)	1918	6	50	
Kahoma	*	3		
Kahana	*	3.8		
Kauaʻula	*	4.5	25.5	upgraded in 1929
Launiupoko	*	0.8		
Olowalu	*	4	11	
Ukumehame	*	3	15	

Sources: C. Wilcox, *Sugar Water,* 1996; O'Shaughnessy, "Irrigation," 1902.
* Date unknown

cement-lined canal; and an inverted siphon that was 3 feet in diameter and spanned 1,253 feet across the ʻĪao Valley. By 1913 Wailuku Sugar Company had two additional ditches taking water from Waiehu Stream to the north of Wailuku, five on Wailuku Stream in the ʻĪao Valley, and two on the Waikapu Stream to the south of Wailuku.[136]

Pioneer Mill Company obtained its irrigation water from two major sources: seven relatively short ditches it developed to bring water from valleys on the western slopes of West Maui plus the Honokōhau Ditch, which was built and owned by Honolua Ranch, located on the northern slopes of West Maui. The original Honokōhau Ditch was surveyed in 1901, begun in 1902, and completed in 1904 (see Table 3.9). The design was to take water at an elevation of about 700 feet from Honokōhau, Kalaunui, and Honolua streams on Honolua Ranch property. With a total length of 53,240 feet, 16,300 feet would be tunnels (6.5 feet high and wide), with the remainder consisting of open ditch, flumes, and five inverted siphons to cross the gulches. The agreement between Pioneer Mill Company and Honolua Ranch was for Pioneer Mill to finance construction of the ditch and purchase up to 15 million gallons of water at a cost of $3,000 per mgd per year for the first five years, and thereafter at a cost of $2,750 per mgd per year. Any flows over 15 mgd would be at no cost to Pioneer Mill. Honolua Ranch would own the ditch, repay construction costs, and maintain the ditch at its own expense.[137]

Soon after it was constructed, however, the ditch began to experience severe problems, including leakage and landslides. Work on a new ditch commenced in 1912; this time, most of the open ditch was replaced by tunnels. In total, the design of the new ditch specified 31 contiguous tunnels separated by horizontal access tunnels (adits) from the side of the mountain to the main tunnel. The total length of tunnels would be 34,241 feet. In addition, the ditch would include 1,183 feet of inverted siphons, 726 feet of covered flumes, and only 427 feet of open ditch. The floors of the tunnels were lined with concrete and the sides were plastered, a decision that was later regretted due to deterioration of the plaster. This second ditch, known as the Honolua Ditch, had a capacity of 50 mgd and was completed (without a fatal accident) in 18 months at a cost of $239,841.[138]

Despite the fact that the new ditch was partially lined with concrete, in less than a decade the seepage losses between the intakes and the Pioneer Mill weir were estimated to be 31 percent. In 1923 Pioneer Mill Company undertook a five-year project to reline the ditch. In order to keep irrigation water coming to the plantation, water was diverted around the section being renovated through the old 1904 Honokōhau Ditch. Pack mules were used to carry cut stone used in lining the tunnels, and small locomotives were used to haul materials inside the tunnels. By the end of the five-year project, the capacity of the ditch had been increased to 70 mgd, and much of the old open ditch could be filled in and the flumes dismantled.[139]

Among Pioneer Mill Company's other ditches, the largest was the Honokōwai Ditch which took water from Honokōwai Stream north of Lahaina and conveyed it southward to the plantation's fields. The original ditch was built in 1898; within 20 years the flume had deteriorated and had to be replaced by a much larger concrete-lined tunnel approximately 6 feet tall by 6 feet wide. A 21-year lease to cross government lands was obtained in 1917. The 1.55-mile-long tunnel was drilled simultaneously from the north and south entrances, with each end using "three shifts of two drillers, two helpers, three muckers, and a shift boss." The Honokōwai Ditch project was completed in 1918. Though its capacity was 50 mgd, the average flow of the ditch was actually about 6.15 mgd due to lack of flow in the Honokōwai stream.[140]

Little information remains about the other six Pioneer Mill ditches. It is known that the original Kauaʻula Ditch was replaced by a 4,013-foot cement-lined tunnel in 1929. Its capacity was 25.5 mgd, and its average flow was 4.5 mgd. This project had 30 workers (11 Filipinos and 19 Japanese), and before its completion an explosion killed a drill operator and a foreman. The Kahoma Ditch had inlets at 1,930 feet and 960 feet, and its median flow was 3.19 mgd. One of its two intakes flowed into an 8-inch pipe, and it had a median flow of 3.68 mgd. Some of its water was used by Maui County, and Lahainaluna School had a right to 4.5 hours per day of its flow. The Launiupoko Ditch had a median flow of only 0.78 mgd. Pioneer Mill Company also drilled a number of water development tunnels to try and tap the high-elevation aquifers in central West Maui, but these efforts produced little additional water.[141]

Olowalu Plantation, with fields to the south of Lahaina, was bought by Pioneer Mill Company in 1930. It had two ditch systems: the Olowalu Ditch, with a capacity of 11 mgd and a median flow of 4.08 mgd; and the Ukumehame Ditch, with a capacity of 15.5 mgd and a median flow of 3.30 mgd.[142]

Summary

For Hawai'i's sugar industry, the period from annexation to the Great Depression was one of rapid growth and consolidation, including improvements in irrigation infrastructure, mechanization of field and factory operations, and development of housing and amenities needed to support the plantation labor force. Development of hybrid varieties of sugarcane by the HSPA, especially H 109, complemented the advances in other aspects of sugar production. H 109 began the replacement of the pure *Saccharum officinarum* varieties that had dominated the industry. The new hybrids were more tolerant of insect predation and disease and demonstrated hybrid vigor; consequently, they substantially increased sugarcane and sugar yield potential. Sugar prices, which averaged 3.3–4.1 cents per pound between 1898 and 1913, were sufficient to support industry expansion. In 1912 the industry could boast 53 sugar plantations, with seven of those on Maui. It had 45,000 employees throughout the Hawaiian Islands in 1917, and its production doubled during the first quarter of the twentieth century. Improvements in mechanization included widespread use of large steam plows, chemical rather than mechanical weed control, and steam railways with both permanent and temporary portable tracks to transport cane from the field to the mill. In addition to railroads, in areas with abundant water at high elevations, "wet flumes" were built to float cane down from fields to the rail lines or to the sugar factory. In a few cases bundles of cane were hooked to trolleys attached to overhead cables in high-elevation fields.

These trolley-borne bundles rapidly descended by gravity to rail lines or the factory far below.

After World War I gasoline and diesel tractors began to replace steam plows, and smaller tractors began to replace horses and mules for lighter work. Major work continued on renovating, rebuilding, and maintaining ditch systems and sealing ditches to minimize leakage. Plantations realized the importance of organic and mineral fertilizers, applying greater and greater amounts, especially ammonium sulfate and Chilean sodium nitrate. They also made use of other sources of plant nutrients, including animal manure from plantation stables and mud removed from cane juice by factory filter presses. Legume manures were also tried for periods when fertilizer prices were high, but these were discontinued when prices dropped. Factory operations improved with larger, more powerful mills, more efficient systems to "clarify" the juice, multistage steam-heated vacuum evaporators to concentrate the juice, and improved centrifuges to separate molasses from the raw sugar crystals. By the early decades of the century, Hawai'i probably had the most technically advanced sugar industry in the world.

Major labor unrest and strikes, primarily by Japanese workers between 1900 and 1920, led to substantial wage increases and improvements in labor conditions. In addition, the HSPA began recruiting large numbers of Filipino immigrants, and by the 1920s Filipinos and Japanese were the largest ethnic groups employed by the industry. To increase efficiency and minimize labor unrest, plantations also began to employ more independent contractors to supply labor for field operations like cultivation, irrigation, and harvesting.

The sugar industry developed rapidly on Maui. Just after the turn of the century, HC&S's factory at Pu'unēnē was the largest and most modern in the world. Plantation owners began to dig "Maui-type" skimming wells to tap the vast reserves of freshwater lying atop seawater under the islands. The large pumps were initially driven by coal- and oil-powered steam engines, which were later replaced by electric motors. Power was supplied by generating electricity from burning excess bagasse and then distributed through a plantation-installed grid. HC&S developed a complex system of irrigation tunnels and ditches to bring water from plantation and territorial lands on the rainy windward side of East Maui. Maui plantations—well aware of the value of irrigation water—conducted major programs to reduce seepage losses by lining their main irrigation ditches. Pioneer Mill Company developed its own ditch system to bring water from the north coast of West Maui to its fields on the dry leeward slopes above Lahaina.

By the 1920s HC&S had 3,000 employees and a plantation population of 7,000 men, women, and children living in over 25 camp communities. The workers were supplied with company housing, improved sanitation, and a variety of community services, including stores, churches, schools, and health services. But the "boom" of the first three decades of the century would soon be tested by two worldwide crises, the Great Depression and World War II.

4

Depression, War, Federal Legislation, Science, and Technology—1930 to 1969

The four decades from 1930 to 1969 brought significant social and political challenges and dramatic technological improvements in field and
factory. The Great Depression, World War II, the Jones-Costigan Act (Sugar
Act), and labor strife all took their toll on Hawai'i's sugarcane industry. Low
sugar prices during the 1930s were followed by severe labor shortages during World War II. These stresses caused the industry, led by scientists at the
Experiment Station of the Hawaiian Sugar Planters' Association (HSPA), to
focus on increasing sugar yields while reducing labor costs. Working closely
with plantation staffs, they engineered some dramatic improvements in sugarcane varieties, crop management, and factory operations, substantially improving labor efficiencies. Seeking efficiencies of scale, sugar companies
merged during this time, reducing the number of plantations on Maui from
six to three and throughout the Hawaiian Islands from 45 to 25, as shown
in Table 4.1.

Statewide, sugarcane acreage harvested declined by 17 percent from
1930 to 1968, largely because cane production was abandoned in some
poorly adapted areas, but also due to domestic sugar quotas established by
the Sugar Act. Nevertheless, total sugar production increased by 32 percent because of a 59 percent increase in tons of sugar produced per acre
harvested. This dramatic increase in sugar production per acre was the
result of the industry's commitment to science and technology. Major
changes included breeding more productive varieties and improving soil
and plant analysis, fertilization, irrigation control, harvesting systems, and
mill operations. Plantation workforces declined due to mechanization of
planting, weed control, irrigation, fertilization, and harvesting. Plantations
converted from animal and steam to diesel and gasoline power for planting
and cultivation. Steam was replaced by electricity and diesel power for
pumping irrigation water. The large teams needed for cutting and loading
cane into railcars and flumes were replaced by mechanical harvesting and

Table 4.1. Number of plantations, harvested acres, tons sugar, and tons sugar per harvested acre on Kaua'i, O'ahu, Maui, and Hawai'i in 1930 and 1968

Island	Plantations 1930	Plantations 1968	Acres harvested 1930	Acres harvested 1968	Tons sugar 1930	Tons sugar 1968	Tons sugar/acre 1930	Tons sugar/acre 1968
Kaua'i	11	8	29,701	25,036	195,133	263,882	6.57	10.54
O'ahu	9	4	24,734	21,860	246,597	224,737	9.97	12.29
Maui	6	3	22,024	18,292	192,408	288,128	8.74	13.18
Hawai'i	19	10	59,677	48,337	296,489	455,435	4.97	9.42
Total	45	25	136,136	113,525	930,627	1,232,182	6.84	10.85

Sources: Gilmore, *Gilmore Hawaii Sugar Manual,* 1931; Bloomquist, *Sugar Manual,* 1969.

loading, and cane transport systems were converted from rail and flume to large trucks.

This chapter describes the industry's evolution during these four decades, including severe stresses caused by federal legislation, low prices during the Great Depression of the 1930s, labor shortages during World War II, and the rapid mechanization and labor unrest that followed the war. It also covers the many improvements made by HC&S and other plantations as they sought to increase yields and reduce costs.

Sugar Production and Prices

The 1930s got off to a good start for sugar growers in Hawai'i, with the industry producing its first million-ton sugar harvest in the 1931–1932 season: an impressive 1,025,354 tons of sugar were produced from 139,743 harvested acres. But sugar prices declined during the Great Depression, falling from a high of 6.94 cents per pound during the boom times of 1920–1924 to a low of 2.57 cents per pound on the New York market in May 1932; yearly average prices remained below 3 cents per pound even as late as 1938, 1939, and 1940. As a result of these stresses, harvested acreage fell from 144,959 in 1933 to 126,116 in 1935 (see Table 4.2). Harvests then fluctuated between 126,000 and 139,000 acres until the beginning of World War II. Despite low sugar prices, the industry was able to maintain sugar yields between 6.92 and 7.97 tons per acre throughout the decade of the 1930s.[1]

Federal legislation (the Jones-Costigan Act, referred to as the Sugar Act of 1934, and amended in 1937 and 1948) severely restricted the domestic sugar industry by establishing domestic sugar production quotas and rules

Table 4.2. Hawaiian sugar production statistics from 1930 to 1965

Year	Cane acreage	Harvested acreage	Tons sugar	Tons sugar/acre
1930	242,761	133,840	939,287	7.02
1935	246,491	126,116	986,849	7.82
1940	235,110	136,417	976,677	7.16
1945	211,331	103,173	821,216	7.96
1946	208,376	84,379	680,073	8.06
1950	220,383	109,405	960,961	8.78
1955	218,819	106,180	1,140,112	10.74
1958	221,683	84,136	764,953	9.09
1960	224,617	103,584	935,744	9.03
1965	235,576	109,600	1,217,667	11.11

Sources: Gilmore, *Gilmore Hawaii Sugar Manual,* 1931; Bloomquist, *Sugar Manual,* 1969.
Note: Industry-wide strikes severely reduced acreage harvested in 1946 and 1958.

for labor compensation and treatment. Quotas were also set for foreign imports and for regional domestic production. Taxes on domestic processing were used to compensate growers for limiting their production to marketing quotas. As the result of a 1936 Supreme Court decision, Congress passed an amended version of the act—the Sugar Act of 1937—separating the excise tax from payments to domestic producers. The 1937 law was extended several times until it was replaced by the Sugar Act of 1948, which changed quota allocations and gave preference to Cuban sugar in return for its cooperation in supplying the United States with additional sugar supplies during World War II. The 1948 act was extended to 1960, when imports from Cuba were suspended as a result of the Cuban revolution. This act remained in effect until the Sugar Act of 1974 was passed.

The Sugar Act was implemented at no cost to taxpayers but was opposed by consumers because it limited the amount of foreign sugar entering the United States, thereby increasing the price. The act was initially opposed by the sugar industry as well, but from the 1930s until the 1970s it benefited the industry by controlling the U.S. supply of sugar, thereby stabilizing prices at levels profitable to the Hawaiian industry. However, the Sugar Act was terminated by Congress in 1974, producing a wild swing in sugar prices, just the event that the act was designed to prevent.[2]

Prices begin to rebound in 1941 (to 3.39 cents per pound) and were maintained at 3.74 cents per pound as the result of government price controls instituted during World War II. However, labor shortages throughout the war resulted in a gradual reduction in cane acreage harvested, declining from 136,417 acres in 1940 to 103,173 acres in 1945. Because of an industry-wide strike in 1946, acreage harvested dropped to a low of 84,379 acres. But after the strike was settled, acreage rebounded and remained quite stable, fluctuating between 108,794 in 1949 and 106,742 in 1957. But sugar yields per acre rose dramatically in the mid-1950s, from 8.78 tons per acre at the start of the decade to over 10 tons per acre every year between 1953 and 1957 (see Table 4.2). Yields per acre decreased for a variety of reasons in the late 1950s, to a minimum of 8.83 tons in 1959, but saw a steady increase to 11.12 tons per acre in 1966 before retreating slightly to 10.85 tons per acre in 1968.[3] These data represent average yields of cane grown for 18 to 24 months for the entire industry, including low-sunlight unirrigated lands. Irrigated lands such as those on Maui produced much higher yields.

Fertilizers

The Experiment Station of the HSPA analyzed both the increases in sugar yields in the early 1950s and their declines after 1955. The yield increases in the first half of the decade were attributed to increased fertilizer use and new varieties developed by the Experiment Station. For example, from 1951 to 1955 the percentage of new varieties increased from 55 percent to 95 percent on irrigated plantations and from 14 percent to 70 percent on unirrigated plantations. In addition, from 1951 to 1957 nitrogen fertilizer rates for irrigated cane almost doubled, from 212 pounds to 421 pounds per acre. Potash fertilizer rates also more than doubled, from 205 pounds per acre in 1951 to 430 pounds in 1957. Fertilizer rates for unirrigated plantations increased almost as much. Much of the increase in nitrogen fertilization was probably the result of economical aqua ammonia becoming available in Hawai'i in 1953. However, analysis of maximum plantation yields between 1950 and 1957 demonstrated that maximum yields were most often found in fields receiving 225–275 pounds nitrogen per acre and 275–375 pounds potash per acre, significantly less than the average of 421 pounds of nitrogen applied to irrigated cane in 1957.[4]

Industry average cane and sugar yields began to decrease in 1956 and continued downward, falling slightly below 9 tons of sugar per acre in 1959. In fact, sugar yields fell proportionally more than cane yields. Multiple factors probably contributed to these lower yields, but the most severe losses could be attributed to a devastating labor strike in 1958 (see below), which effectively stopped irrigation and resulted in the harvesting of overage cane in the 1958, 1959, and 1960 crops. Another likely factor was the probably excessive and late application of fertilizer nitrogen, which stimulates cane growth but can reduce the percentage of sugar in the cane.[5]

Hawaiian producers generally paid more attention to increasing cane yields than the sugar content of the cane. They could do this because the fields and mills were under the same ownership. If the mills had been independent of the growers, there would have been more interest in growing cane of higher quality because the mills would have insisted on grinding cane of the highest possible quality to maximize their profits.

Labor

Inadequate availability of labor and the need to reduce labor costs were nearly constant issues from the 1930s through the 1960s. In the early 1920s Japanese outnumbered Filipinos on the islands' sugar plantations; however, as a result of active recruitment in the Philippines by HSPA, by 1930 nearly 70 percent of the industry's (almost 50,000) adult male workers were Filipinos, and only 18.5 percent were Japanese. The final HSPA-sponsored group of immigrants from the Philippines arrived in 1946 and included about 6,000 men and over 400 women and 900 children. Thousands more Filipinos came to Hawaiʻi on their own after the HSPA program ended, mostly between 1965 and 1975 after the U.S. Immigration Act of 1965 eliminated restrictions on the number of Asians who could enter the United States.[6]

In the late 1930s—prior to the U.S. involvement in World War II—the cash wages paid by the Hawaiian sugar industry (not counting benefits like housing and medical care) were the highest paid by any sugar industry worldwide. For example, the U.S. Department of Agriculture set the base wage of $2.27 for Hawaiʻi's sugar workers, compared with $1.50 per day in Louisiana. Average daily wages compiled by HSPA for May 1938 were as follows: cane cutters $2.27; cane loaders, $2.61; portable railroad track handlers, $2.83; and irrigators, $1.85.[7]

Workers in the industry were normally paid under one of three systems: as day laborers, short-term contractors, or long-term contractors. Day laborers were simply paid for the number of hours they worked, and they were managed by plantation *lunas,* or supervisors (see Figure 4.1). Short-term

Figure 4.1. Luna on horse at harvest. Photo courtesy of HARC

contract work, also called "piece work," referred primarily to field work paid each pay period and calculated by the amount of work that had been accomplished, such as how much cane was cut or loaded by an individual or group of workers. By the early 1930s piece work was becoming more common in the industry. It provided "a powerful incentive to efficient, persistent, faithful effort, because earnings depend upon the units of work performed." Loading railroad cane cars by hand and piling cane to be loaded by mechanical means were examples of individual piece work because each worker could be assigned a specific car that would be weighed at the mill. Cutting cane by hand was considered group piece work, because a group of cutters would progress down the rows together, and the tonnage of cane from an area cut by the group could be calculated from the weight of cane in the railcars loaded from that area.[8]

Some typical rates of pay for piece work were: cutting burned cane (20 cents per ton), cutting green cane (27 cents per ton), loading cane by hand (25 cents per ton), piling cane for a loading machine (18 cents per ton), and operating a loading machine (4 cents per ton). In addition, plantations commonly paid premium rates to more productive workers. For example, loading burned cane by hand paid 25 cents per ton for the first 90 tons in a month, with the rate per ton increasing to 26 cents per ton for tons 130 to 149 and 27 cents per ton for tons 190 to 199. Members of crews doing group piece work could receive different rates of pay depending on their individual responsibilities. For example, tractor drivers and men who rode the plow to keep it free of trash had different rates of pay per unit of land cultivated.[9]

Long-term contract work referred to work such as irrigating, fertilizing, weeding, and otherwise tending specific fields. The rate of pay was negotiated on the basis of the production of cane on a specific field, and it could vary depending on the productivity of the soil, the variety of cane, and the difficulty of cultivating or irrigating the land. Typically, a labor contractor would negotiate a price per ton of clean cane harvested from a field. He would then employ a crew of workers to provide cultivation services, usually about one person per 10 acres. The plantation would advance payment monthly based on a conservative estimate of the yield, and upon harvest of the field, the contractor would receive the difference between the total advanced and the value of the cane harvested. He would then distribute that difference among the members of his crew. Thus, his final payment depended on the quality of work and its effect on final yield. Laborers who left the crew before harvest would receive only the pay advanced monthly and would forfeit their share of the difference. Typical rates for long-term contract work of cultivating, fertilizing, and irrigating fields were as follows: $1.10 per ton for plant cane (up to 7.5 acres per worker) and $1.15 for ratoon cane (up to 8.5 acres per worker).[10]

Regular full-time hours per week for day laborers varied greatly among plantations and among jobs on plantations. For example, in 1929 the typical workweek was 10 hours per day, six days a week, with the sixth day often shorter than the other five. Some plantations, however, worked nine hours per day, and many factory employees worked 12 hours a day six days a week for a 72-hour work week. Of course, contract laborers, especially those doing long-term contract work, had considerable flexibility in the hours they worked, and many even brought their wives and children to the field to assist them.[11]

Labor turnover was high on plantations, with workers coming and going for many reasons. In 1929 the annual rate of turnover was slightly over

30 percent for men and about 75 percent for women. In true paternalistic fashion, plantation owners were concerned that "our Filipinos are a restless lot, changing around from place to place." This was thought to be "due to the fact that Filipinos have relatives in great numbers and . . . they want to be with a cousin, uncle, or brother, or some other connection" on another plantation.[12]

Plantation field work was nearly always strenuous, sometimes dangerous, and few workers could be on the job every workday month after month. In 1929 a plantation reported that its adult male employees worked an average of 88 percent of the workdays. To discourage absenteeism, plantations paid a bonus of 10 percent if an employee worked at least 23 days in the month, and in 1929 slightly over 70 percent of the employees typically collected the bonus.[13]

In 1935 the U.S. Congress passed the National Labor Relations Act, and in 1937 the Supreme Court declared it constitutional, paving the way for workers to organize and pursue collective bargaining with management. In Hawai'i at that time, two-thirds of the islands' sugarcane workforce was Filipino, but they were the lowest paid of the ethnic groups. In 1937 about 3,500 Filipino workers on Maui struck four plantations for higher wages and dismissal of several supervisors. After 85 days, the HSPA approved a 15 percent pay increase, and the strikers returned to work.[14]

By the late 1930s the International Longshore and Warehouse Union (ILWU) was actively organizing in Hawai'i, but World War II brought wage and job freezes, halting any movement toward unionization of sugar workers. As the war drew to a close, the 1945 Hawai'i Employment Relations Act made it clear that agricultural workers could unionize, and by 1946 the ILWU had obtained contracts to represent sugar, pineapple, and dock workers. Some 21,000 sugar workers on 33 plantations went on strike that September. The strike lasted 79 days, during which time irrigation ceased. The work stoppage, combined with reduced planting during the last years of World War II, greatly decreased acreage harvested and sugar produced in 1946. The strike ended with the industry compromising with workers for wage increases, and at the workers' request many of their benefits, such as plantation housing, were converted to cash wages.

By the mid-1950s mechanization had dramatically reduced the workforce in the sugar industry. With the ILWU contract up for renewal in 1958, the union began making preparations for a possible strike as early as 1954, setting up a strike fund and educating members. The industry responded with its own educational program to gain public support. The ILWU pointed out that the industry's labor force had decreased by 26 percent between 1947 and 1954, and in roughly the same time frame the cost of labor had declined from 36 percent to 22 percent of industry revenues. When negotiations began in earnest, the ILWU requested a 25 cent per hour increase in wages, plus reclassification of many jobs that required exceptional skills. The industry countered with a much lower wage offer, and so the strike began. The union was well prepared with its strike fund, and it put its members to work on community improvement projects to garner public support. In addition, it paid some workers to continue to irrigate cane fields to reduce losses. After 128 days a new contract was signed giving the workers a 25 cent wage increase over the course of the three-year contract, and the union claimed victory.[15]

Research and Development

The Great Depression had a severe effect on the sugarcane industry and its 36 plantations in Hawai'i. Average New York sugar prices had fallen from a high of 6.94 cents per pound during the boom times of 1920–1924, to 4.28 cents per pound in 1925–1929, 3.17 cents per pound in 1930–1934, and 3.24 cents per pound in 1935–1939. In response to the financial crisis, plantations requested that their financial support of the Experiment Station of the HSPA be reduced. Subsequently, some Experiment Station programs were streamlined, and staff was reduced from 55 in 1932 to 46 in 1935; however, as the industry began to recover, station staff numbers rose to 66 in 1941. But the onset of World War II caused major changes in Experiment Station activities, with several staff leaving for active military duty, the pathologists producing penicillin for civilian use and yeasts for island bakeries, entomologists helping with mosquito control, and agriculturists assisting in production of food crops needed to feed the islands' residents. When the war ended, staffing rebounded, and by 1945 the Experiment Station had grown to 75 scientists and support staff.[16]

Breeding

The Experiment Station cane breeding program was very active in the 1950s. Breeding goals included finding new cane varieties that (1) did not flower in response to short days in the winter, (2) lodged without damage to stalks or root systems, (3) tolerated cool winter temperatures, (4) continued to grow actively in the second year, (5) had good juice quality, (6) produced heavy tonnages of sound cane at harvest, (7) ratooned well, (8) resisted diseases, and (9) had good milling characteristics. In order to expand the germplasm base of the breeding program, Experiment Station scientists collected sugarcane varieties and sugarcane's wild relatives in New Guinea. These were brought back to Hawai'i and crossed with commercial varieties to increase desired traits such as disease resistance and vigorous growth. Variety testing stations were maintained on the four major islands, and both preliminary and replicated variety trials were installed every year on plantations.[17]

Crop Protection

Diseases, insects, rats, and weeds were continuing concerns for sugar growers. In the early 1950s the principal diseases were chlorotic streak, leaf scald, and pineapple disease. By the end of the decade, red rot, ratoon stunting, and *Pythium* root rot had also become significant. Although Fiji disease, smut, rust, and downy mildew had not been found in Hawai'i, the likelihood of their eventual introduction prompted evaluation of the resistance of Hawaiian varieties in Fiji, where the diseases were endemic. Various methods of control were also evaluated. For example, hot water treatment of seed cane was found to be effective in controlling chlorotic streak and ratoon stunting disease. Adding phenyl mercuric acetate to the hot water seed treatment controlled pineapple disease. The only practical control for red rot, leaf scald, downy mildew, and Fiji disease was thought to be varietal resistance.[18]

Prominent insect pests included the anomala beetle, though the parasitoid scoliid wasp was providing relatively good control. The beetle borer was increasing on all islands, probably because softer cane varieties were being developed, but the tachinid parasitoid fly was expected to provide adequate control. The mealybug was of concern because it could be a disease vector, but its effects on cane production were unclear. Entomologists feared the introduction of other insect pests, especially leafhoppers, which were vectors for Fiji disease. In response, an early warning system consisting of blacklight traps was installed near the Honolulu airport and Pearl Harbor to detect any unwelcome insect pests arriving by air or sea.[19]

The Norway rat was an important pest that had not been controlled by introducing the mongoose, but baits containing Warfarin, Endrin, and thallous sulfate had been developed and were being distributed by air or in bait dispensers.[20]

Weed control was, of course, a continuing challenge. Prior to 1950 the three major herbicides were 2,4-D (2,4-dichlorophenoxyacetic acid) for post-

Figure 4.2. Applying herbicide by helicopter. Photo courtesy of HARC

emergence control of broadleaf weeds, CADE with PCP (concentrated activated diesel emulsion with pentachlorophenol) for contact weed control, and STCA (sodium trichloroacetate) for grass weeds. In the early 1950s CMU (3-(p-chlorophenyl)-1,1-dimethylurea) and DCMU (3-(3,4-dichlorophenyl)-1,1-dimethylurea) became available for pre-emergence use, and AR-CADE (an aromatic oil version of CADE) became available as a contact herbicide. In the mid-1950s Dalapon (2,2-dichloropropionic acid) and Silvex (2,(2,4,5-trichlorophenoxy)propionic acid) became available, with Silvex used for post-emergence weed control of brush species and dalapon (Dowpon) for control of rhizomatous grasses. By the early 1950s herbicides were being applied by helicopters, spray planes, spray tractors, and workers with knapsack sprayers; and research on improved tanks, spray nozzles, field vehicles, and herbicide formulations and mixes continued through the decade (see Figure 4.2). Herbicide injury to cane was found in some soils with little ability to adsorb the herbicides and prevent uptake by the cane roots. Some dalapon-tolerant strains of weeds such as Bermuda grass were also identified.[21]

Plant Physiology

Beginning in the 1930s and continuing through the 1960s, the Experiment Station conducted a great deal of research on basic plant physiology to understand the factors affecting cane growth and development. Studies evaluated the effects of day length, drought stress, and chemicals on tasseling, and breeders began to control temperatures and day length to induce flowering and facilitate crossing. Air-conditioned greenhouses were used to demonstrate the relative importance of light energy, air temperatures, and root temperatures on cane growth, nutrient uptake, and sugar production. For example, root temperatures below 62° F reduced water consumption, nitrogen uptake, translocation, and growth, even when air temperatures were optimal. At air temperatures above 74° F, light was the principal factor limiting cane growth, and cooler air temperatures favored sugar accumulation over cane growth, explaining why juice quality was better during periods of cool temperatures.[22]

Planting

Experiment Station engineers also worked closely with plantations to reduce the costs of field operations like planting. By the 1950s a single machine could open furrows 10–16 inches deep, apply a balanced fertilizer in the bottom of the furrow (and sometimes inject ammonia), cover the fertilizer with a small amount of soil, drop in seed pieces 12–24 inches long (ideally with three buds or "eyes"), and cover the seed pieces with an inch or two of soil, all in one operation.[23]

Soil Compaction

In the 1950s and 1960s soil compaction, caused principally by mechanized cane harvesting, was a growing concern. Experiment Station scientists found that cane roots were often unable to penetrate compact layers below the depth of tillage, and if they did, the roots were deformed and probably ineffective in absorbing water and nutrients.[24]

Since cane was normally ratooned two or three times before replanting, it was important to correct any compaction problems before replanting. Typical plantation practice was to pull a three-tine subsoiler at a depth of 18 inches with a D8 Caterpillar tractor. The subsoiler tines were often equipped with attachments to lift and shatter the soils, and the subsoiler was normally pulled across the slope to create deep channels that encouraged infiltration of water. Cross-subsoiling at 45° or 90° angles to the first pass was recommended to avoid leaving strips of compacted soil between the tines. After subsoiling, heavy plows with 44-inch to 52-inch discs were able to plow to depths of 16–20 inches. A disc harrow was then used to break up clods and produce enough small soil aggregates for a good seedbed. After harrowing, irrigation furrows could be laid out with the assurance that several inches of uncompacted soil would be below the young root system, encouraging vigorous root development. In general, irrigated plantations like HC&S, Pioneer Mill Company, and Wailuku Sugar Company had fewer compaction problems than unirrigated plantations because irrigation ceased 30 to 45 days prior to harvest, usually leaving soils dry and resistant to compaction at harvest.[25]

Heavy cane haulers ("Tournahaulers" and tractor-trailers) were normally restricted to in-field roadways to limit soil compaction; however, these roadways often required substantial restoration in order to avoid stunting subsequent cane growth. Many plantations, including HC&S and Pioneer Mill Company, spread cane trash and excess bagasse from the mill on these areas and tilled it in. This practice increased infiltration of irrigation water, stimulated soil microbial activity, and improved cane growth. HC&S and Wailuku Sugar Company used similar applications of cane trash and bagasse to improve root growth and yields on the sandy soils on the isthmus between East and West Maui.[26]

Soil Fertility

Experiment Station scientists made important advances in sugarcane fertility and soil management from the 1930s through the mid-1960s.[27] H. F. Clements, working at the University of Hawai'i through the 1970s, developed the "crop log" method of repeatedly sampling the nutrient and moisture contents of specific tissues throughout the crop and using the data to manage fertility, irrigation, and ripening.[28]

By the 1930s Experiment Station scientists had demonstrated that cane varieties differed, sometimes substantially, in their uptake and optimum tissue concentrations of major nutrients: nitrogen (N), phosphorus (P), potassium (K), calcium (Ca), magnesium (Mg), and silicon (Si).[29] In the 1950s visual symptoms of nutrient deficiencies and toxicities were demonstrated by growing cane in carefully controlled nutrient solutions.[30] These observations were extended to micronutrients by 1960.[31]

Soil sampling and plant analysis formed the basis for routine plantation nutrient management programs that identified fertility problems and corrected them with applications of fertilizers, lime, and organic soil amendments. These advances in understanding soil fertility allowed plantation staff to make preliminary diagnoses of nutritional problems in the field, verify them with soil and plant analyses, and correct problems economically and confidently. As a result of nuclear reactor developments during World War II, radioisotopes like N^{15}, rubidium $(Rb)^{86}$, P^{32}, Ca^{45}, K^{42}, and S^{34} became available for agricultural research. Research with these isotopes demonstrated that potassium, phosphorus, and sulfur are rapidly translocated within the plant, with calcium moving much more slowly. Experiment Station scientists learned that only one-third or less of the fertilizer nitrogen applied to the crop is actually taken up during the year that it is applied. Most of the remainder is absorbed by soil microorganisms, and only becomes available over time. In addition, about half the fertilizer nitrogen absorbed by the crop is returned to the soil in the leaves and tops of the plant—parts that are not taken to the mill at harvest. In addition to laboratory and greenhouse studies, Experiment Station scientists conducted hundreds of field experiments in cooperation with plantations. Results of the field tests led to an enormous increase in fertilizer use: from 1923 to the mid-1960s annual nitrogen fertilizer consumption by the industry rose from 9,000 to 20,000 tons, phosphate from 3,000 to 9,500 tons, and potash from 2,000 to 20,000 tons.[32]

Prior to 1952 ammonium sulfate was the principal form of fertilizer nitrogen, but aqua ammonia (20 to 21 percent nitrogen) became available in Hawai'i in 1953. In 1955 R. P. Humbert, a soil scientist at the Experiment Station, reviewed 584 replicated nitrogen fertilizer experiments in Hawai'i and concluded that nitrate and ammonium sources of nitrogen were equally effective. Aqua ammonia was quickly adopted by the industry because it was a more economical source of nitrogen, and it could be applied either in irrigation water or injected into the soil to minimize volatilization.[33]

The first application of nitrogen in irrigation water in the Hawaiian sugar industry probably occurred in 1902 at Kīhei Sugar Company on Maui; sodium nitrate was dissolved in a 50-gallon barrel of water and then allowed to flow into the main irrigation canal, reducing the cost of labor required to apply fertilizer by hand. The labor-saving advantages of applying fertilizer in irrigation water were still important in the 1950s, and numerous studies showed that the ammonia remained in the top three inches of soil in the irrigation furrow, though poor water distribution along the row caused uneven distribution of the nitrogen (see Figure 4.3).[34]

In addition to application via irrigation water, by the 1960s aqua ammonia was being applied directly on or under the seed pieces at planting and/or injected into the soil on both sides of the young cane in plant and ratoon

Figure 4.3. Injecting aqua ammonia into irrigation canal. Photo courtesy of HARC

crops. Around the same time liquid and granular fertilizers combining nitrogen, phosphorus, and potassium became available. Equipment was built to apply bulk liquid and solid materials, greatly reducing the labor needed to handle sacks of fertilizers.[35]

Leaching losses of nitrogen and potassium were minimized when fertilizers were applied to fields on which cane was growing because the cane roots and microbes quickly absorbed the nutrients. Little leaching of phosphorus occurred in most Hawaiian soils because it was tightly adsorbed by clay particles. But calcium could move downward in acid soils, especially when provided as gypsum rather than lime. Still, this was beneficial because it reduced aluminum toxicity in deeper soil layers and could improve root system development.[36]

Field research repeatedly showed that adequate nitrogen fertilizer increased total cane and (usually) sugar yields. Young cane plants had the ability to take up "luxury" amounts of nitrogen, store it in the stalk, and later translocate it to the growing leaf and stem tissues. Early application of much of the nitrogen was important to development and survival of first-year suckers, but application of too much nitrogen late in the crop often stimulated late sucker growth, reducing sugar concentrations in the harvested cane. Therefore, for "short" crops that would be harvested around one year of age, almost all the nitrogen was applied in the first few months. For "long" crops harvested at about two years of age, half to two-thirds of the nitrogen was applied at the start of the crop. By the 1950s plantations were applying the last application no later than 10 months before harvest.[37]

Sugarcane also requires large amounts of potassium for rapid growth. However, most of the potassium absorbed by the crop is stored in the stalk and is extracted along with the sugar when the cane is crushed in the mill. This potassium ends up in the molasses, which is normally sold for animal feed. Unless the plantation soils are rich in potassium, the element must be replaced by fertilizer potassium. Humbert calculated the difference between potash (K_2O) purchased by plantations and potash sold as a constituent of molasses between 1930 and 1957. For the entire industry, a total of 64,905 tons more potash left the plantations in molasses than was purchased in fertilizers. Of the four sugar-producing islands, only Hawai'i purchased more potash than it exported in molasses (60,736 tons), and Maui had the largest deficit (72,476 tons). Every year from 1923 to 1952, for instance, Wailuku Sugar Company exported more potash in molasses than it purchased as fertilizer. Inevitably, these deficits in potash fertilization began to show up as declines in soil potash. In 1938 one-quarter of the fields in the Waihe'e section of the plantation north of Wailuku were low in potassium, but by 1951 over three-quarters were low.[38]

Concern over potassium fertility led to further study. Of the 432 potash experiments conducted between 1940 and 1954, 35 percent produced significant increases in cane yields. Virtually no response to potassium fertilization was observed when extractable soil potassium was greater than 325 pounds per acre-foot (ac-ft) of soil, but in virtually all cases cane responded positively to fertilizer potassium if extractable soil potassium measured less than 200 pounds per ac-ft.[39]

Phosphorus fertility was also a concern, especially in highly leached acid soils. These soils—and especially their subsoils—had little plant-available phosphorus, and fertilizer phosphorus was readily "fixed" in unavailable forms. During the early 1900s some plantations had noted positive responses of cane growth to phosphate-containing fertilizers, and from the late 1920s to the mid-1930s large amounts of phosphate fertilizers were imported and applied to these highly leached acid soils, mostly in the high rainfall areas of the unirrigated plantations. Between 1940 and 1954 the industry conducted a total of 354 phosphorus fertilizer rate experiments, of which 57 (16 percent) produced statistically significant responses in sugar yields. Little or no yield response was found when available soil phosphorus was greater than 65 pounds per ac-ft, but significant responses could be expected if levels were lower than 35 pounds per ac-ft. It is important to note that only 28 of the 354 experiments were conducted on Maui, and only one of those produced statistically significant differences in sugar yield per acre. After the 1940s phosphorus nutrition was not a great concern on Maui plantations.[40]

Experiment Station research demonstrated that adequate soil calcium reduces aluminum and manganese toxicity, reduces leaching of potassium, increases availability of soil phosphorus, and increases juice quality. But little or no response to calcium application could be expected if soil exchangeable calcium was greater than 400 pounds per ac-ft.[41]

Aerial fertilizer application also became important in the 1950s and 1960s. In response to studies conducted by the Pineapple Research Institute, the pineapple industry began large-scale aerial aqua urea applications in 1949, and the sugar industry began tests the following year. Detailed physiological research confirmed that the urea is easily absorbed through the leaf stomates, quickly converted to ammonium and then to organic nitrogen compounds that are translocated to parts of the plant with low nitrogen concentrations. Visual responses of the cane canopy were evident as yellow, nitrogen-deficient leaves quickly greened up. Tests at numerous plantations, including HC&S and Wailuku on Maui, supported aerial application of nitrogen; however, later research demonstrated that application after 12 months of age often resulted in poor juice quality.[42]

Studies with radioactive rubidium, which mimics potassium in soils and plants, along with the results of traditional field experiments have suggested that aerial application of potassium is much more effective than soil application. Measurable increases in plant potassium levels and recognizable visual effects of aerial potassium fertilization were evident as leaves became more green and turgid. In addition, leaf tips were intact, not frayed as they were in potassium-deficient plants; however, the effects were usually short-lived because only a small amount of the nutrient was absorbed.[43]

Twenty thousand acres of cane in the islands had received some fertilizer by air (see Figure 4.4) as of 1952. After that, the availability of potassium fertilizers increased, and potassium chloride was added to the aerial mix. Initially, aqua urea and aqueous potassium chloride were applied at rates of 40 pounds of nitrogen and 19 pounds of potash in 10 gallons of water per acre. However, beginning in 1953, researchers realized that aerial application of granular fertilizers also produced good crop responses. Fertilizer pellets that fell into the crop canopy lodged at the base of the leaf blade, where dew or small amounts of rainfall dissolved them and allowed

Figure 4.4. Applying liquid fertilizer with biplane. Photo courtesy of HARC

efficient uptake of the nitrogen and potassium. Larger amounts of rain simply washed the nutrients into the soil, allowing uptake by the root system. Plantations began to aerially apply 200–300 pounds per acre of granular materials. In 1954, roughly 5 million pounds of fertilizers were applied aerially, and from 1956 to 1958 about 20 million pounds were applied by airplane each year, mostly on unirrigated plantations after the cane closed in.[44]

Irrigation

By the 1930s Maui's plantations had developed extensive ditch systems to collect surface (mountain) water from the high rainfall areas of East and West Maui and move it to their cane fields. They had also dug a number of wells to tap both high elevation and basal aquifers. These ditch systems were capable of supplying large amounts of irrigation water. But losses on the plantations—due both to leakage from secondary ditches moving the water from the main ditches to the fields and poor distribution within the fields themselves—were major concerns. In addition, the industry had little understanding of how sugarcane water use varied among locations, seasons of the year, and ages of the crop.

Maui's sugar companies were leaders in all aspects of improving irrigation efficiency and effectiveness in the four decades from 1930 to 1969. They continued to improve the ditch systems that harvested water from East and West Maui and brought it to the plantations. Industry leaders drilled more wells, installed more powerful and efficient pumps, lined plantation ditches, and developed less labor-intensive in-field irrigation systems. Finally, HC&S took a lead role in working with Experiment Station scientists to determine just how much water sugarcane needed for maximum growth.

As labor became more expensive and difficult to obtain, Experiment Station scientists worked with plantation irrigation staff to develop less labor-intensive systems.[45] If intervals between irrigations could be lengthened without reducing crop yields, both labor expenses and water could be saved. Attempts to improve interval control began in the 1930s, when U. K. Das, an Experiment Station meteorologist, proposed the use of day-degrees as a guide to irrigation. However, a number of field experiments were unsuccessful in establishing a link between temperature and the onset of crop water stress, primarily because soils differed in their water holding capacity, and in Hawai'i solar radiation, not temperature, largely determines the rate of crop water use. The industry then turned to soil moisture, plant tissue moisture, and the rate of cane elongation as indicators of when to irrigate. At about the same time, L. D. Baver, director of the Experiment Station, proposed a meteorological approach to estimating crop water use.[46]

Soil Moisture Sensors and Water Use

During the 1950s the sugarcane industry used two types of soil moisture sensors to routinely schedule irrigation in sugarcane fields. Bouyoucos blocks buried in the soil below the cane stool worked by monitoring the electrical resistance between electrodes embedded in the body of a plaster of paris block. As water was extracted from the soil by the cane root system, the moisture content of both the soil and the block decreased, causing an increase in resistance between the block's electrodes. By correlating the block's resistance with the elongation rate of the youngest cane leaf, it was possible to predict from resistance measurements when cane growth would begin to experience moisture stress. However, the relationships among block resistance, soil moisture tension, and cane growth varied with the depth and water holding capacity of the soil, the depth of the block in the soil, the temperature and salinity of the soil, and the age of the block. As a result, the manufacture, calibration, and installation of the blocks were standardized, and by 1957 they were in use on eight plantations covering 41,686 acres.[47]

The tensiometer was another instrument perfected during 1950s for irrigation control. The tensiometer is a fairly simple design—a tube filled with water—that measures the relative ease or difficulty a plant's roots will experience as they attempt to draw moisture from the soil. One end of the tube is a porous ceramic cup that allows water to move in and out in response to soil moisture tension. The other end has a gauge to measure the tension of the water in the tube, up to a maximum of 1 atmosphere. When the soil is moistened by irrigation or rainfall, water moves from the soil into the tensiometer and the gauge reading decreases to near zero. As the soil surrounding the porous cup dries, water moves out of the tensiometer into the soil, and the gauge reading increases. With proper correlation with cane growth rates, tensiometers are good indicators of soil drying and its effects on cane elongation. In the 1950s tensiometers were in use on 27,000 acres at HC&S, where they were placed in every field at depths of 18–24 inches in locations two-thirds of the way down the row, normally the first area in the field to begin suffering water stress. When the tensiometers began to register changes in soil moisture, it meant that the cane roots had extracted most of the available water in the top 18–24 inches, and it was time to irrigate.[48]

In general, experiments indicated that near-maximum cane and sugar yields were obtained when irrigation was applied at tensiometer readings of approximately –0.25 atmospheres or at Bouyoucos block readings of approximately 5,000 ohms. These readings typically corresponded to removal of approximately 50–60 percent of the available water in the zone of active root growth. In 1953, 13 of 17 irrigated plantations in Hawai'i were using Bouyoucos blocks, and HC&S was using tensiometers to schedule irrigation. This careful monitoring of soil moisture and crop growth allowed plantations to reduce excessive irrigation amounts that had been the norm in earlier decades.[49] For example, by 1958 crops at Māʻalaea, Pūlehu, and Kāheka, Maui were irrigated at rates of 0.56–0.58 cm (0.22–0.23 in) per day. This was a dramatic decrease from the 1889 crops at Spreckelsville and Hāmākuapoko, Maui, which were estimated to have received 1.01–1.23 cm (0.40–0.48 in) per day. It was also much less than the average of 0.89 cm (0.35 in) per day applied to fields at Wailuku Sugar Company and the 1.22 cm (0.48 in) per day applied to HC&S cane from 1912 to 1916.[50]

Research also revealed the importance of using relatively large plots (approaching one acre) for irrigation experiments. Subsurface water movement from fields above the experiment and lateral root growth between experimental plots could mask the effects of irrigation treatments applied to small plots.[51]

In the late 1950s and early 1960s Experiment Station scientists, in cooperation with HC&S staff, conducted several important field experiments to understand the microclimate of sugarcane and its effects on water use, soil moisture extraction, and irrigation requirements. In the Evapotranspiration Project, begun in 1957, cane was planted in large metal boxes 5 feet wide, 8.7 feet long, and 4 feet deep sunk into three fields at HC&S. These boxes, called "drainage lysimeters," provided detailed measurements of weather conditions, the amounts of irrigation water applied, rates of evaporation from a U.S. Weather Bureau Class-A evaporation pan, drainage from the boxes, and growth of the cane. This experiment provided valuable basic information on the effects of weather and crop age on water use and growth of sugarcane. They demonstrated that over periods of a few weeks to a month, evaporation from Class-A evaporation pans correlated well with water used by the cane growing in lysimeters. In addition, the Penman evaporation equation, which uses temperature, wind speed, humidity, and solar radiation to estimate evaporation from an open water surface, correlated well with pan evaporation.[52]

By the late 1960s the in-field drainage lysimeters used in the HC&S Evapotranspiration Project were supplemented by hydraulic weighing lysimeters, which produced accurate measurements of crop water use over periods of one to a few days.[53] These lysimeter studies revealed that water loss from standard Class-A evaporation pans installed at ground level within cane fields provided a good estimate of evapotranspiration by a well-watered sugarcane crop. Based on these data, Experiment Station scientists recommended that plantation personnel begin to compute the daily water balance of their crops and use that water balance to guide the date—and, if possible, the amount—of irrigation needed by individual fields. Such studies led to the widespread use of standard Class-A pans, sometimes in conjunction with Bouyoucos blocks or tensiometers, to schedule irrigation before soil water was depleted to the point that cane growth began to suffer.[54]

These lysimeter and pan evaporation studies were supplemented by another experiment at HC&S known as MASI, an acronym for method, amount, surface shape, and interval of irrigation. This study demonstrated that during periods of drought stress, cane roots can extract significant amounts of water from 2–5 feet below the soil surface, even though root density at these depths is very small and roots are often confined to vertical fissures in the soil.[55]

A review of industry irrigation research in the 1950s and early 1960s[56] concluded that:

- Pan evaporation and sugarcane water use (evapotranspiration) are highly correlated with solar radiation.
- Well-watered sugarcane used about 40 percent of Class-A pan evaporation when the crop was very young. This percentage increased to about 110 percent of pan evaporation when the cane leaf canopy fully covered the soil.
- During the winter, water use rates in the HC&S experiments were about 0.19 inches (0.48 cm) per day, but they increased to about 0.34 inches (0.86 cm) per day in summer, for an annual average of 0.22 inches (0.56 cm) per day (70–90 inches per year).
- When gypsum block resistance increased to about 5,000 ohms in the zone of maximum root concentration (6–8 inches), irrigation should be applied. This corresponded to removal of about two-thirds of the available soil moisture by the crop. Yield loss began to occur when the crop was allowed to extract more than 67–75 percent of the available moisture in the top 6–8 inches of soil.
- When, as at HC&S, tensiometers were installed at a depth of 18 inches, irrigation should be applied when the tensiometer read −0.25 atmospheres—when about 60 percent of soil water had been used by the crop.
- In sandy soils, tensiometers were good indicators of soil water contents. Bouyoucos blocks were more reliable in clayey soils.
- For furrow-irrigated cane, elongation of the youngest leaf (an indicator of stalk growth) and photosynthesis continued normally until soil moisture in the zone of maximum root development approached the wilting point, then both almost ceased. After irrigation, both elongation and photosynthesis recovered almost completely in one or two days.
- Under furrow irrigation, infiltration of irrigation water varied greatly among soils, decreased as soil moisture increased, decreased greatly in compacted soils, increased with the age of the crop, and decreased with increasing slope of the furrows.
- Under sprinkler irrigation, infiltration increased as surface cover (cane trash or leaf canopy) increased. Infiltration was greater when the soil was left flat than when it was deeply furrowed. Infiltration decreased as slope increased and as the energy of rainfall increased.

- Plantations generally needed to reduce furrow slopes (for example, from 2.5 to 0.5 percent), increase rates of flow from the flumes into the furrows (for example, from 18 to 25 gallons per minute), and minimize clogging and misalignment of flume outlets.

Subsequent reanalysis of several irrigation experiments on Oʻahu and at HC&S revealed that relative cane yield (actual cane yield divided by maximum cane yield in the experiment) increased linearly until relative water use (actual water use divided by maximum water use in the experiment) reached 1.0. In the furrow-irrigated treatment of the MASI experiment on HC&S, water use efficiency varied from about 1.01–1.12 tons cane per acre-inch of "effective water" (water transpired by the crop), up to about 138 inches. In a parallel sprinkler-irrigated treatment, cane yields of effective water were slightly less, ranging from about 0.88–1.01 tons cane per acre-inch of effective water. In these two experiments on HC&S, regardless of whether furrow or sprinkler irrigation had been used, the time at which drought stress occurred had little effect on the relation between water use and cane yield.[57]

Furrow Irrigation

By the 1950s the Hawaiian sugar industry had been using furrow irrigation for over a century, with an irrigator able to irrigate 6–20 acres per day. However, with traditional irrigation methods, excessive amounts of water often infiltrated in the first half of the furrow because the water was flowing over that section for a longer period of time than the last half of the furrow. This excess water often moved down beyond the root zone of the crop and could eventually recharge the basal aquifer or move downslope within the field, emerging as a problematic seep. In contrast, less water than the cane needed often infiltrated into the soil in the lower half of the furrow because flow was inadequate, especially after the growing stools of cane began to fill up the furrow and impede water flow. Finally, the water that did not infiltrate often accumulated at the end of the row, producing another area of excessive irrigation. Industry-wide, furrow irrigation efficiencies were estimated to vary from 30–50 percent; however, irrigation engineers recognized that these estimated efficiencies were, "at best, approximations" due to the large number of variables affecting infiltration.[58] These included rooting depth, soil moisture-holding capacity, soil infiltration rate, furrow length, furrow slope, flow rate, flow time, and cane age. Plantation irrigation staff faced formidable problems in designing and operating

Figure 4.5. Herringbone irrigation system on a steep slope. Photo courtesy of HARC

irrigation systems on Hawaiian soils, which could vary by a factor of eight in their rates of irrigation water intake. As a result, optimum furrow lengths varied from 150–1,000 feet, and optimum flow rates into furrows varied from 10–100 gallons per minute.[59]

Nevertheless, continuing pressure to reduce labor costs and improve irrigation water use efficiency led to innovation. By the 1960s at least five types of furrow irrigation were in use.[60]

- *Herringbone irrigation.* A permanent flume of precast concrete or aluminum was normally installed running down the slope with furrows running away from the flume on the contour so that water could be distributed to furrows on either side of the flume. The system got its name from the appearance from the air of a fish skeleton with the flume as the spine and the furrows as the bones diverging at angles (see Figure 4.5).
- *Continuous long-line irrigation.* Temporary aluminum (or sometimes galvanized iron or plastic) flumes were laid down the slope on top of and approximately perpendicular to the furrows, which were laid out on the contour. The system allowed water to be discharged from several flumes into continuous furrows up to 1,000 feet long running under the flumes. The system allowed for low flow rates from the flumes

Figure 4.6. Continuous long-line irrigation system. Photo
courtesy of HARC

and long set times required by soils with low infiltration rates. Flumes
were removed before harvest and replaced after planting the next crop
(see Figure 4.6).[61]

- *Level ditch irrigation*. An irrigation water supply ditch was run on the
contour along the slope and fed level furrows that ran down a slight
slope to the next supply ditch. Furrows connected these level ditches
so that water that left the end of the furrow without infiltrating entered the next level ditch down the slope to water the furrows below
it. The system was best adapted to relatively level lands because the distance between level ditches decreased as the slope of the field increased
(see Figure 4.7).[62]

- *Semiautomatic furrow irrigation*. Semiautomatic systems were introduced in 1965 to reduce irrigation labor requirements. In a typical system, control gates were installed at intervals within ditches in order to
convey water down the slope of the field. When an irrigation round
began, all gates were open, allowing water to flow to the lowest section of the field, where it flowed into the furrows. After a predetermined time that allowed the lowest section to be irrigated, a timer closed
the control gate below the next lowest section of the field, allowing
water to flow into furrows in that section for a predetermined time.

Figure 4.7. Level ditch irrigation system. Photo courtesy of HARC

The process was repeated until all sections of the field had been irrigated. Gates and timers had to be reset between irrigation rounds, but the system reduced irrigation labor requirements to approximately one-third of the manual irrigation method. Semiautomatic irrigation was quickly accepted and by the end of 1967 had been installed on approximately 5,500 acres.

- *Automatic furrow irrigation.* Continuing shortages and high costs of labor led to the first completely automatic furrow irrigation system, which was installed in Maui in 1966. It used hydraulically actuated gates that opened when water reached them, allowing water to enter the furrows. The control gate then closed when water reached a predetermined point in the furrow, stopping flow into the furrows. By closing the gate, water was allowed to flow to the next control gate, which opened when water reached it. Water pressure supplied by a pump through small plastic tubes was used to operate the gates.

Sprinkler Irrigation

By the late 1960s Experiment Station engineers were working with plantation irrigation staff to evaluate several types of sprinkler irrigation systems.

- Low-volume, closely spaced sprinklers had sprinkler heads elevated above the height of the crop at spacings of 60–90 feet.

Figure 4.8. Large-volume sprinkler irrigation. Photo courtesy of HARC

- Portable, large-volume sprinklers were mounted on trailers. They took water from ditches along plantation roads and used pumps to spray at rates of up to 3,000 gallons per minute, rotating at about one revolution per minute (see Figure 4.8).
- Large-volume gravity-head systems took water from pressurized buried main irrigation pipes. Portable, quick-couple aluminum pipes were attached to valves and carried the water to large-volume rotating sprinklers. But two crews were required, one to move pipe and another to move sprinklers.
- Hose-reel systems were similar to gravity-head systems and used the same sprinkler connected to an 8-inch-diameter flexible hose wound on a reel mounted on a small tractor. It had higher capital cost, but less labor was required to move pipe.
- Center-pivot, hydraulically driven boom systems used booms up to 1,600 feet long supported on towers approximately 100 feet apart. Each boom system irrigated up to 150 acres, with about 40 hours required for one revolution of the boom.[63]

Despite the research on sprinkler irrigation, furrow irrigation remained predominant throughout the 1960s; as late as 1972 only 3,560 acres of solid-set sprinklers were in use. In typical furrow irrigation systems, water was applied every 10–14 days during the "boom" stage of growth—from about 6–20 months of age.[64]

Harvesting

Until the late 1930s all cane in the Hawaiian sugar industry, like other industries around the world, was harvested by hand. But the majority of the Hawaiian cane had grown for approximately 24 months prior to harvest. Consequently, harvest crews were faced with a tangled mass of recumbent cane that had grown tall, fallen over (lodged), and turned upward again, usually lodging more than once. This mix of intertwined vertical and horizontal stalks, most alive but some dead, was "a forest of tangled stalks 10 to 25 feet in length, and often in excess of 100 tons per acre."[65]

To burn a field of cane, a backfire was first set on the downwind side of the field. After it had burned a short distance into the field, the remaining borders of the field were lit. Vandercook reported that cane fires were "spectacular," with flames roaring and the smoke rising "mountain high," and "after a good dry weather burn, the tall cane stalks are left almost clean and apparently quite undamaged"[66] (see Figure 4.9).

After the field was burned, workers (one man to a row) used large cane knives to lop off unburned tops, cut the stalks as close as possible to the soil surface, and cut the long stalks into convenient lengths. Loaders stacked the stalks into piles weighing about 75 pounds each and loaded them onto carts or railcars for transportation to the mill or, increasingly, the stalks were stacked in windrows for mechanical loading. In the 1930s some cane in high-elevation fields was still loaded into flumes and floated down to the mill or to a rail transfer station where it fell directly into cane cars for transport to the mill. To increase recovery of sugar at the mill, operations were highly coordinated to minimize the time between cutting and milling.[67]

Railroads had been adopted by plantations for the transport of cane beginning in the 1870s, and by 1938 there were 32 plantation railroads with 900 miles of permanent track, 300 miles of portable track, and 140 locomotives. Seven independent railroads served island communities but also transported sugar (see Figure 4.10).[68] Always looking to improve, plantations began evaluating different types of trucks and associated trailers to replace their railroads. Though World War II interrupted the transition, by the late 1940s and 1950s large trucks, special cane-hauling trailers, and vehicles like the enormous Tournahauler began to replace the narrow-

Figure 4.9. Cane fire smoke. Photo courtesy of Lance Santo

gauge railways and flumes that had long been used to transport cane (see Figure 4.11).

Labor costs put serious pressure on the plantations to reduce hand labor, starting with loading the hand-cut cane. By the 1930s some plantations were beginning to use mobile derricks with "grabs" to lift piles of cane onto railcars. But the grabs could also be used to simply tear the burned but uncut "forest of tangled stalks," from the soil and lift the mass directly onto railcars, greatly reducing the labor costs of harvesting. In 1938 'Ewa Plantation Company on O'ahu became the first plantation to use grabs for essentially its entire harvest. While achieving the goal of reducing labor, the grab-harvested cane carried to the mill all the cane tops that would normally

Figure 4.10. Steam engine pulling cane cars across trestle. Photo courtesy of HARC

Figure 4.11. Loading cane onto Tournahauler with dragline.

have been removed, as well as large amounts of soil, and often rocks, discarded metal, and other damaging trash. This extraneous material could account for a fourth of the total weight carried to the mill, and, of course, reduced the efficiency of sugar extraction and damaged mill machinery.[69]

An improvement to the grab harvest system was developed at Waialua Agricultural Sugar Company. The "Waialua rake" was "simply a heavy steel pipe contraption in the form of a huge, half-closed hand." Two tractors with steel cables wound around drum attachments dragged the rake back and forth across the field, breaking the stalks off at the roots and leaving the cane in piles for the loaders "complete, as in the case of the grab harvester, with trash, mud, and rocks." The rake was "fast and cheap to operate," but the mills complained of the "unsorted rubble" that was brought in with the cane.[70] Despite these efforts to mechanize harvesting, in 1939 "the great majority of Hawaiian fields continue[d] to be harvested by the old hand method."[71]

The next step toward mechanization was the "push rake," essentially a Waialua rake mounted in place of the blade on a D-6 Caterpillar or similar tractor. As the tractor was driven into the tangled mass of recumbent cane, the teeth lifted the cane off the ground and the frame of the push rake broke the cane stalks off at ground level, though it also pulled some of the stools out of the ground with roots and soil attached (see Figure 4.12). The operator pushed the cane into windrows about 50–100 feet apart. The push rake was then followed by a light spring-tooth *liliko* rake mounted behind a tractor to gather any cane missed by the push rake. A large

Figure 4.12. Harvesting cane with push rake. Photo courtesy of HARC

Figure 4.13. Harvest operations at HC&S. Photo courtesy of HARC

dragline-mounted (or later, hydraulic) grab then picked up the windrowed cane and loaded it onto tractor-trailers for transport to the mill (see Figure 4.13). While push rakes improved the harvest process, large amounts of trash and extraneous material were still carried to the mill. This problem increased in wet weather, when even more soil and trash from green or poorly burned cane reached the mill.[72]

Because harvest and hauling accounted for over one-fourth of the cost of cane production, and because the harvest methods used for recumbent two-year cane substantially reduced sugar recovery at the mill, in the 1940s, 1950s, and 1960s a number of harvesting and field-transport machines were developed and tested. These machines were especially intended for use on unirrigated lands where conventional harvesting with push rakes

Figure 4.14. V-cutter. Photo courtesy of Lance Santo

and grabs under wet conditions was difficult. The most successful of these unconventional harvesting approaches were the "V-cutter" and the "cane buggy."[73]

The V-cutter was a D-6 Caterpillar or similar tractor with an 8-foot-wide V-shaped blade much like a snowplow (see Figure 4.14). At the tip of the blade a large revolving vertical cutter was mounted. The lower edge of the blade was lowered just above the soil and, as the tractor moved forward, the vertical cutter sliced through the tangle of recumbent cane; the V-shaped blade divided the mass into windrows about 6–8 feet apart, which could then be lifted onto cane buggies or tractor-trailers with a grab loader. By the mid-1950s the V-cutter was in widespread use, especially on unirrigated plantations. Cane buggies were self-propelled in-field transport units with wide tracks that were loaded with eight to 10 tons of cane. The buggies transported the cane to the edge of the field and dumped it for subsequent loading onto tractor-trailers for transport to the mill. The industry developed and tested other less successful approaches, including a single machine for cutting, loading, and transporting cane to the edge of the field, where it could be loaded onto tractor-trailers. Other unconventional machinery included transportable field-cleaning units that used air blasts to remove trash from the cane before it was loaded onto tractor-trailers. Though the industry had spent millions working to perfect harvesting and hauling equipment, in 1968 Humbert concluded that progress had been "slow and costly."[74]

Figure 4.15. Puʻunēnē mill. Photo courtesy of HARC

Factory Operations

During the 1930s plantation factories varied in size, with the smallest mills grinding about 10,000 tons of cane a year, and the largest, at Lihuʻe Sugar Plantation on Kauaʻi, capable of grinding 600,000 tons of cane and producing 70,000 tons of sugar per year. However, the mill with the greatest sugar production was HC&S's Puʻunēnē mill (see Figure 4.15), which could produce 80,000 tons of sugar, though because of its better quality cane, it actually milled less cane than Lihuʻe.[75]

Vandercook gives a good account of factory operations in the late 1930s.[76] Factories had developed different methods to remove cane trash, soil, and other extraneous material before the cane reached the shredding knives, crushers, and rollers of the mill. "One [unnamed] farm has installed a costly mechanism that subjects the dirty cane to a winnowing, blowing process. On another the cane rolls down an incline of steel rollers beneath powerful jets of water. At Kīlauea on Kauaʻi, the cane moves through a narrow channel of running water. At an estate on Maui the stalks are bumped and jostled down a long incline under a blast of air. At Pioneer Plantation they send the harvest through a tank of mixed molasses and water so combined that the specific gravity of the cane permits it to float while heavier matter sinks to the bottom." Despite these steps, according to Vandercook, the "additional soil, grit, and rocks produce[d] extra wear on the mill rolls, conveyors, pumps, and pipes, thus greatly increasing maintenance costs."

After being "cleaned" to the extent possible, the cane passed up a conveyor through "a series of flailing knives" that chopped and shredded the stalks into "a coarse, pulverized mass ready for the crusher." The crusher was typically "two large rolls with immense, interlocking corrugated teeth" that expressed 50–60 percent of the juice. The mat of shredded cane that exited the crusher entered the mills—9–18 rolls, usually about 34 inches in diameter and 78 inches long, arranged in groups of three in the form of a triangle, one above and two below. As the cane passed in sequence through the three to six sets of 3-roll mills, the remaining juice was expressed. In a final extraction process called "maceration," hot water was sprayed onto the moist mat of shredded cane just before the last set of three rollers to extract as much as possible of the remaining sugar before it was also squeezed out of the mat. The bagasse exiting the last set of rollers was dry enough that it could be moved directly to the mill's boilers where it was used as fuel to generate steam to power mill operations and produce electricity.[77]

The juice expressed by the crusher and mills was pumped to lime tanks, where it was heated to stop fermentation, and a solution of lime was added. The juice was then pumped to the clarifier tanks, where it remained for several hours as the lime caused small particles of impurities to coagulate and settle as sediment to the bottom. The clear juice at the top of the tanks was pumped to the evaporators, and the sediment was pumped to filter presses or centrifuges to separate the solids from the liquids in the sediment.[78]

The clarified juice, now consisting of about 85 percent water and 15 percent sugars, was then pumped to a series of vacuum evaporators. In a typical quadruple evaporator system the juice in the first evaporator was boiled at just over the boiling point of pure water (212° F) with steam generated in the factory boilers. The slightly concentrated juice then passed to the second evaporator, where the steam exiting the first evaporator kept it hot, but pumps pulled a vacuum in the atmosphere above the liquid. This vacuum caused the liquid in the second evaporator to boil at a lower temperature (about 203° F). The same process was followed in the third and fourth evaporators, with steam from the previous evaporator keeping the juice hot and increased vacuum causing it to boil at successively lower temperatures.

After passing through the quadruple evaporator system, the concentrated juice was at a temperature of about 130° F and contained about 65 percent sugars and 35 percent water. The low-pressure, low-temperature steam leaving the fourth evaporator was cooled and condensed with a spray of cold water, then pumped out of the system, helping maintain the vacuum.[79]

The concentrated juice then passed to the vacuum pans, "large iron or copper cylinders standing in a vertical position, with dome-like tops and conical bottoms," where the juice was finally boiled to the point at which sugar crystals formed. Like the evaporators, the vacuum pan was under a strong vacuum and was heated by steam. As the concentrated juice boiled and became even more concentrated, the sugar boiler watched the process through a small window in the pan. Eventually, sugar crystals began to grow, while the water with its minerals and organic impurities remained in the liquid phase. As the crystallization process proceeded, the sugar boiler carefully added more concentrated juice to replace the water that was boiling away. Eventually, the contents of the pan became *massecuite,* or "cooked mass," a French term for the viscous mixture of sucrose crystals and molasses.[80]

The massecuite was then passed to the centrifuge, invented in 1853 in Honolulu by David Weston. The centrifuges used in the 1930s were cylindrical, about 40 inches in diameter and 2–3 feet deep, with perforated brass baskets lined with fine-mesh brass screens. The massecuite was loaded into the centrifuge, which was then spun at high speed to allow the liquid molasses—but not the sugar crystals—to pass through the screen. After molasses had been centrifuged out of the mass, the pale brown sugar crystals that remained were scooped out by hand for bagging and export. Hawaiian sugar was shipped to the mainland or other markets in 105-pound jute bags. Plantation owners continued to evaluate labor-saving mechanical dischargers to remove the sugar from the centrifuges after the molasses had been expelled. In the 1830s typical Hawaiian sugar was 97.5 percent pure sugar, greater than the world standard of 96 percent.[81]

A great deal of research was conducted in the 1950s on by-products of sugarcane production. Molasses, bagasse, and cane trash could also be used as major constituents of feeds for beef cattle, swine, and poultry. Molasses was also used to produce yeast protein, but it could not compete economically with soybean meal. Bagasse could be used in the production of corrugated cardboard, ceiling boards, bleached pulp, newsprint, hardboard, furfural, and lightweight structural concrete. In addition, excellent quality wax could be collected from mill cane wash water, and fatty acids and alco-

Table 4.3. Total pol losses in the Maui sugar mills in 1930 and 1968

Mill	1930	1968	Source of increased losses
	percent		
HC&S (Puʻunēnē)	8.5	14.9	bagasse and molasses
MAC (Paia)	8.3	13.6	bagasse and molasses
Wailuku Sugar Company	10.8	15.0	bagasse
Pioneer Mill Company	10.0	13.4	molasses
Olowalu	13.16	—	
Kaeluku	20.45	—	

Sources: Gilmore, *Gilmore Hawaii Sugar Manual,* 1931; Bloomquist, *Sugar Manual,* 1969.

hols could be extracted from mill filter cake; unfortunately, with the exception of Canec, a ceiling board, no commercially viable applications were found.[82]

Research in the late 1940s and 1950s revealed that 10–13 percent losses of sugar occurred during harvest, transport to the mill, and cleaning at the factory prior to crushing and milling. Most of these losses resulted from mechanical harvesting of cane, which required installation of complex cane cleaning systems to reduce the rocks, soil, and leaf trash entering the mills. Despite these systems, extraneous materials continued to enter the crusher with the cane, reducing the efficiency of sugar extraction and juice clarification. An estimate from the 1950s put the cost of inefficiencies caused by milling dirty cane at more than $11 per ton of sugar produced.[83] Table 4.3 gives the total losses of sugar in 1930 and 1968 from Maui's sugar mills, demonstrating the effects of mechanical harvesting on the recovery of sugar in the factory.

Maui's Groundwater Resources

The sugar industry has developed two major sources of groundwater on Maui. "High-level" water is trapped in an unusual area of permeable lava flows within a complex of vertical, intrusive, and impermeable basalt dikes on West Maui. The dikes inhibit both downward and lateral flow of water to the basal aquifer, and the maximum water level is indicated by springs. These occur from about 700–1,200 feet at the edges of the dike complex up to about 3,500 feet near its center. These aquifers are unusual because

East Maui and the other Hawaiian Islands have radial rift complexes that are less effective impediments to downward water movement.[84]

West Maui's high-level springs were very productive where streams cut into the high-level aquifer, and even during low-flow periods about 9 million gallons per day (mgd) issued from several important high-level springs. Sixteen tunnels with a total length of 28,666 feet were dug to tap the high-level aquifer on West Maui between the years 1900 and 1926. These tunnels produced about 20.5 mgd, or 715 gallons per day per foot of tunnel. However, a number of the tunnels tapped water that would otherwise have issued as spring flows.[85]

Basal aquifers were the most important sources of groundwater developed by the sugar industry on Maui, and their characteristics were well understood by the 1940s. The basal aquifer under West Maui floats atop seawater in a coastal belt that extends inland until it reaches the island's large central rift zone and dike complex. Leakage from the outermost dikes of the central high-level aquifer, percolation of rainfall, and infiltration from streams all contribute to the recharge of Maui's basal aquifers. In addition, leaking irrigation ditches and overirrigation of sugarcane and pineapple fields contributed to the aquifer. [86]

The top of the basal aquifer in the middle of the isthmus between the East and West Maui volcanoes was only about 5.5 feet above sea level, and water levels in irrigation wells within the basal aquifer were known to be remarkably stable. They fluctuated, though only slightly, in response to tides, rainfall, and pumping and seepage of irrigation water from ditches, reservoirs, and fields. On an annual basis, the highest water levels were typically found in the spring before pumping started and were lowest in the late fall before it ceased, but these annual fluctuations were usually only 0.5–1 feet. During drought years water tables were slightly lower than in wet years. For example, the average water level in HC&S wells was 10 inches lower in 1935, a drought year, than in normal years. The water level in observation wells near large pumps responded within a few minutes when pumping started or stopped, but level increases due to recharge from rainfall normally took several days.[87]

The salinity of wells could vary from week to week, month to month, and year to year due to variation in pumping and recharge. During the dry 1935 irrigation season, the mean salinity of one of HC&S's wells increased from about 65 grains per gallon (gpg)—equivalent to 1,114 milligrams per liter (mg/l)—at the beginning of irrigation in May to about 95 gpg (1,628 mg/l) in late September, when pumping was reduced as a result of rainfall. Salinity then declined to 80–90 gpg (1,371–1,542 mg/l) for the remainder of the season. In the wetter year of 1936 the average salinity of the same well remained at 65–85 gpg (1,114–1,457 mg/l) and in 1937 between 55–70 gpg (943–1,200 mg/l), gradually increasing both years between the beginning of irrigation in the winter or spring and its end in the late fall. Similar results were found for West Maui.[88]

An inventory of groundwater conducted in the 1940s estimated the quantity of discharges from springs on East and West Maui. It revealed large amounts of undeveloped groundwater, principally from the basal aquifers; for instance, about 100 mgd of freshwater per day were being discharged to the sea from the northern half of West Maui. The study suggested that much of the water could be captured by locating more Maui-type wells a mile or two inland in the northern part of West Maui.[89] On the northeastern shores of East Maui, Honomanū Springs—which discharged from two sites at an elevation of about 15 feet at the mouth of Honomanū Valley—produced about 5 mgd. Similarly, numerous small springs issued from sandy beaches between Pā'ia and Kahului between high and low tide lines, producing an estimated 2 mgd. In fact, horses and mules could sense these flows just below the surface of a beach east of Kaupō and were sometimes seen pawing shallow holes in the gravel to reach the freshwater. The study estimated that almost 700 mgd of undeveloped groundwater were available on Maui, most of which was discharged to the ocean from springs near sea level on the windward side of the island (see Table 4.4).[90] It was difficult to access this water because it would have to be raised several hundred feet to the elevation of ditches that could move it to the irrigated sugarcane fields on the drier leeward sides of East and West Maui.

Table 4.4. Average daily discharges of groundwater and rainfall on Maui

	East Maui	West Maui
	(Millions of gallons/day)	
Perched springs	25.0	9.19
Spring-fed streams	74.00	55.50
Tunnels	6.00	20.50
Wells	125.00	45.00
Undeveloped groundwater	205.00	121.00
Rainfall discharges	2,360.00	580.00

Source: Stearns and Macdonald, *Geology and Ground-Water Resources,* 1942.

Maui's Plantations

This chapter has taken a close look at the major operational changes made by Maui's sugar plantations from 1930 to 1969, including large investments in labor-saving technologies. While a significant increase in labor productivity resulted from these changes, smaller gains were made in the industry's cane and sugar yields.

In 1930 six plantations were operating on Maui. Ranked here in order of sugar production from largest to smallest yields, they were: (1) HC&S Company at Puʻunēnē (72,500 tons); (2) Pioneer Mill Company at Lahaina (47,327 tons); (3) Maui Agricultural Company (MAC) at Pāʻia (46,015 tons); (4) Wailuku Sugar Company at Wailuku (18,279 tons); (5) Kaʻelekū Plantation Company near Hāna (5,352 tons); and (6) Oluwalu Company at Oluwalu (2,967 tons). In 1931 Oluwalu Company, which had been organized in 1881 on the southwest coast of West Maui, was purchased and absorbed by Pioneer Mill, its neighbor to the north. Kaʻelekū Plantation Company, which had been organized in 1905, closed in 1945. In order to make better use of East Maui irrigation water and gain other efficiencies, MAC merged with HC&S in 1948.[91] Therefore, after 1945 only four sugar companies were operating on Maui: Wailuku Sugar Company, Pioneer Mill Company, MAC, and HC&S.

Wailuku Sugar Company

Formed in 1862 in northern Central Maui, the plantation produced 800 tons of sugar on 500 acres in 1867. It expanded by purchasing Bal & Adams plantation in 1827, Bailey Brothers in 1884, and Waikapū Sugar Company and Waiheʻe Sugar Company in 1894. Its water-powered mill was later replaced by a steam-powered six-roller mill, and a narrow gauge railroad was installed to replace the ox carts that had been used to bring cane from the fields to the mill. By the turn of the century it was producing almost 8,000 tons of sugar from nearly 1,100 acres harvested. Wailuku acreage and sugar production had more than doubled by 1910 but varied little for the next two decades, with 15,000–20,000 tons of sugar produced each year from 2,200–2,700 acres harvested.[92]

To stimulate rainfall on the plantation's east-facing slopes, the company "started setting out young trees to more thickly cover the mountainsides." In the late 1920s and early 1930s, 20,000 to 40,000 young trees were planted annually, but by 1939 the tree planting program had been abandoned due to declining company profits.[93]

As of 1931 Wailuku Sugar Company was producing cane on 4,481 acres and owned 19,153 acres of forest, pasture, and wastelands. The company increased its harvested acreage by almost 50 percent during the 1930s, but labor shortages during World War II forced it to reduce its harvest, and harvested acreage leveled off to 2,200–2,500 acres between 1955 and 1968. However, tons of sugar per acre increased dramatically as the company increased its fertilizer inputs and mechanization, rising from just under 6 tons per acre in 1945 to 13 tons per acre in 1955. After the plantation recovered from the labor unrest of the late 1950s, yields returned to more than 12 tons of sugar per acre in the late 1960s.[94]

Wailuku had no wells in 1931; all its land was irrigated with mountain water collected from four gulches, the most important of which rose in the ʻĪao Valley and provided 45 percent of the total water in the plantation's ditch system. The gulches provided substantially more water than the 45 mgd normally used to irrigate its cane, and the company was able to sell the equivalent of 10 hours of water per day to HC&S.[95] The company's irrigation infrastructure included 14.1 miles of concrete-lined main ditches, 14.4 miles of wooden flumes, 1,800 feet of 36-inch metal diversion pipe, 700 feet of 26-inch metal diversion pipe, eight tunnels to gather and divert water from high-level aquifers, and 16 reservoirs with total capacity of 71 million gallons to store flows during the night. In the late 1920s Wailuku lined two to three miles of main ditch each year, completing all main ditches by 1932. The next year it began "annually installing several miles of concrete slab for the field ditches." By 1939 over 85 percent of the fields had been changed from contour to long-line irrigation systems, which facilitated mechanical land preparation, aided cultivation, and produced better irrigation results.[96]

After World War II the plantation dug an inclined shaft at a 30° angle 758 feet into the side of the West Maui volcano, allowing it to deliver water from high-level aquifers to the Kama or Waiheʻe ditches at 400 feet and 485 feet above sea level, respectively. The well was used to supplement mountain ditch water during the dry summer months.[97]

Having begun improving its in-field ditches in 1933, the plantation installed 11 miles of concrete flumes over the next two decades. Taking advantage of readily available aluminum, it fabricated 17 miles of semicircular aluminum flumes 11 feet 3 inches long with diameters of 11–30 inches.

Flume sections were laid on top of the cane rows and were lapped 3 inches at each end. Water outlets 4.5 inches in diameter were punched at 5.5 foot intervals.[98] While the aluminum flumes were economical, they had to be removed from the fields prior to harvest and replaced after planting. The labor required to manage both the aluminum flumes and long-line irrigation layout induced the company to convert almost completely to concrete flumes and a herringbone layout by 1969. In addition, the plantation had installed three new pumps in its 758-foot inclined-shaft well, boosting well capacity to 22.5 mgd.[99]

Pioneer Mill Company

Founded in 1862 by Benjamin Pittman, Pioneer Mill Company near Lahaina was one of the oldest plantations in the Hawaiian Islands. Acquired by H. Hackfeld & Co. in 1885 and American Factors in 1918, the plantation grew rapidly from 525 acres and 2,132 tons produced in 1895 to 3,773 acres and 27,229 tons in 1910 and 5,237 acres and 47,327 tons in 1930. In 1931 Pioneer Mill Company gained another 1,200 acres when it purchased Olowalu Company, which had been formed in 1881 on lands of the old West Maui Plantation. By 1935 it was producing cane on over 10,000 acres— half owned and half leased—in a strip over 10 miles long and 1.5 miles wide. Its fields ranged in altitude from near sea level to 2,500 feet, and they were watered by a combination of wells and ditches, the most important of which was the great Honokōhau Ditch.

In its continuing effort to reduce labor costs, the factory began to implement mechanical harvesting to replace manual cutting and loading. This led in the late 1930s to the need to install a cane cleaner. Pioneer Mill Company's harvested acreage declined slightly during the Great Depression, recovered by 1940, and declined again during the labor and fertilizer shortages of World War II. The plantation's strategy for downsizing during World War II was to eliminate "hand work" areas that could not be farmed mechanically, and from 1940 to 1946 the area harvested fell from 5,794 acres to 2,963 acres, reducing sugar production from 47,870 tons to 27,185 tons. In an effort to increase mechanization of field operations in the post–World War II era, the company conducted a major project to remove rocks from over 3,000 acres of cane fields. This allowed trucks into the fields, and by 1951 two-thirds of the cane was transported to the factory by truck.[100] Within four years, harvested acreage had increased to 4,689 and, with exception of the strike year 1958, remained near this level through 1968.[101]

All of Pioneer's cane acreage was irrigated in 1930, with approximately 55 percent of the water supplied by the plantation ditch system and the remainder supplied by wells. Irrigation infrastructure included 18 miles of concrete-lined ditches, a half mile of ditches lined with cut stone, four mountain tunnels (6 feet wide and 6 feet high) with a total length of 10 miles, a little over a mile of water development tunnels (4 feet wide and 6 feet high), one-tenth of a mile of concrete pipe, just under a mile of steel siphons, and 3 miles of metal flumes. The plantation also had five pumping stations with aquifer-level development tunnels and 20 storage reservoirs for a total storage capacity of 235 million gallons. Irrigation amounts ranged from 4–10 inches per irrigation at intervals of 10–30 days, with the last irrigation applied two months prior to the field's expected harvest date.[102]

Between 1930 and 1939 the length of concrete-lined ditches at the Pioneer Mill Company was extended to 22 miles, and cut stone–lined ditches grew to 5.5 miles. Fifty-six percent of the plantation's irrigation water (13,690 million gallons per year) was now supplied by wells with an average lift of 265 feet and a maximum lift of 535 feet. The number of pumping station water development tunnels had been increased from five with a total length of 0.72 miles in 1930 to eight with a length of 1.74 miles in 1939. The number of reservoirs had increased from 20 to 29 since 1930.[103]

Overall, the plantation had made "tremendous strides" in "field layouts and distribution systems," installing "concrete pipes . . . fixed permanently in long lines that travel down the sloping fields . . . [and] fitted at both sides, at intervals that exactly correspond to the width of furrows, with gates closed by metal slides." Irrigators had only to lift the slides to water a furrow and lower the slide when enough water had been applied (see Figure 4.16). These "herringbone" designs allowed two flumes to provide water from both ends of a furrow.[104] These improved field layouts increased the efficiency of plantation irrigation staff from 1.25 acres per day "under the old furrow method" to 10–14 acres per day in 1939.[105]

By 1951 the irrigation system extended 14 miles along the coast and served fields up to 2.5 miles inland up the slopes of West Maui Mountain. Mountain water provided 68 percent and well water 32 percent of the supply for the 8,624 acres of cane. Well water was supplied by nine pump stations with a capacity of up to 25 mgd. In addition, two booster stations were used to raise water to higher elevation fields than the primary pumps were able to reach. All of the pump stations could discharge water into plantation reservoirs, allowing them to operate 24 hours per day. Two of the pump stations

Figure 4.16. In-field concrete irrigation pipes used by Pioneer Mill Company. Photo courtesy of HARC

accessed groundwater through shafts "driven at an incline" down to the basal aquifer, which became thicker inland from the coast.[106]

The main irrigation ditch transmission system had not been improved greatly since 1938, but the plantation continued to make improvements in its in-field systems, which, in 1950, included 155 miles of concrete pipes that delivered water directly to the cane rows. Only 15 percent of the fields still used the level ditch system.[107]

Plantation cane fields stretched 17.5 miles along the coast of West Maui in 1968. The entire cane acreage of 9,389 acres was irrigated, with 45 percent of the water over the previous six years from wells and the remaining 55 percent from mountain water. The total amount of water used in 1965 was 35,382 million gallons, 78 percent of which was used for irrigation. The plantation could pump well water to a maximum height of 759 feet, but 4,500 acres of cane above that elevation were completely dependent on mountain water. Few changes had occurred in the main plantation ditch system, with the exception of adding four reservoirs, for a total of 33 with a total capacity of 252 million gallons, up from 232 million gallons in 1930.[108]

In-field improvements included adding 22 miles of lightweight aluminum flumes. However, no significant improvement in irrigator efficiency had been achieved (9 acres per worker per day). Because of "an increasing labor shortage," the plantation continued to experiment with sprinkler irrigation and installed four types of automated furrow irrigation systems on 558 acres.[109]

Maui Agricultural Company (MAC)

Incorporated in 1921, MAC absorbed the operations of its seven subsidiary corporations, as described in Chapter 3. It occupied a strip of land 2.5 × 10 miles (25 square miles), "extending from a district at sea level with a northerly exposure and an average annual undependable rainfall of about 50 inches; around the shoulder of the [East Maui] mountain to a large district 1,100 feet above sea level with a westerly exposure and an average annual rainfall of about 14 inches, occurring mostly in a few heavy kona storms."[110] The MAC, in addition to growing cane, had extensive acreages of pineapple fields and a ranch with a herd of 1,355 cattle and 163 horses and mules. In the late 1920s the company had about 2,500 acres in pineapples, producing about 20,000 tons of fruit per year, all of which was canned in Kahului by the California Packing Corporation.[111] The plantation also had a forestry department that planted about 50,000 (mostly) eucalyptus trees annually during the 1920s and 1930s, though beginning in 1937 the rate of planting decreased.[112]

By 1930 the plantation had a total of about 9,000 acres in cane and it harvested 4,598 acres that year. Acreage harvested grew to 5,656 acres in 1932, but government cropping restrictions later in the Great Depression caused harvests to decline by over 600 acres. Sugar yields per acre also suffered greatly from drought in 1933 and 1934 and from a severe infestation of eyespot, a fungal disease, in 1938.[113]

The statistics for 1930 reveal that about 5,000 acres were irrigated with mountain ditch water and about 4,000 acres were irrigated with well water. Almost all of the ditch water was provided by East Maui Irrigation Company, which was owned by and supplied water to both the HC&S and MAC. MAC applied an average of about 140 inches of irrigation water during the two-year cropping period in 30 irrigations at 15- to 20-day intervals.[114]

Beginning in 1934, MAC changed from the contour to the long-line irrigation system, producing substantial savings in both labor and water required for irrigation. The normal annual irrigation water supply consisted of 91 mgd of mountain water and 31 mgd of well water in 1938, a significant daily increase over the 104 mgd in 1931. This water was delivered to the fields through 18 miles of masonry-lined ditches, 1.1 miles of tunnels, 7.5 miles of large steel pipes, 4.5 miles of concrete-lined field ditches, 2 miles of concrete irrigation flumes, 8 miles of concrete irrigation pipe, as well as many miles of unlined irrigation ditches.[115]

In 1948 MAC merged with HC&S, bringing under a single management the 22,855 acres of land then under cultivation by both plantations.

Hawaiian Commercial & Sugar Company

As described in Chapter 3, HC&S was founded by Claus Spreckels in 1876, shortly after ratification of the Reciprocity Treaty. It was formally organized as HC&S in 1882, with Spreckels holding the dominant interest. Spreckels invested in its land, irrigation system, factory, harbors, and railroads, but he later landed in financial trouble, and in 1898 H. P. Baldwin, Samuel T. Alexander, and their associates gained control of the company. Table 4.5 gives data on HC&S production from 1895 to 1968.

Cane Acreage and Varieties

In the 1929–1930 sugarcane season, HC&S began grinding its cane on November 19, 1929, and ended on July 19, 1930. It had 14,429 acres planted

Table 4.5. HC&S acreage harvest and sugar production from 1895 through 1968, in five-year increments

Year	Acreage harvested	Tons sugar	Tons sugar/acre
1895	3,290	10,456	3.18
1900	5,556	35,677	6.42
1905	8,929	61,543	6.89
1910	10,519	86,160	8.19
1915	10,686	96,400	9.02
1920	11,144	83,246	7.47
1925	11,182	108,326	9.69
1930	12,203	118,516	9.71
1935	12,262	133,957	10.92
1940	12,674	121,467	9.58
1945	10,200	97,497	9.56
1950	11,119	142,482	12.81
1955	13,276	171,927	12.95
1960	13,408	143,440	10.70
1965	14,166	185,634	13.10
1968	14,679	198,536	13.52

Source: Bloomquist, *Sugar Manual,* 1969.

to cane and harvested 2,056 acres of plant cane, 4,538 acres of ratoon cane, and 1,011 acres of short ratoons, yielding 9.92, 9.79, and 7.62 tons of sugar per acre, respectively. The principal cane variety was H 109, which, since its introduction in 1916, had almost totally replaced Lahaina and D 1135. During the mid-1930s HC&S was forced by government acreage control programs to decrease the amount of cane under cultivation by about 2,200 acres, reducing sugar production for the 1935, 1936, and 1937 crops. But by 1939 the acreage harvested had rebounded to 7,720.[116]

As mentioned earlier, MAC was merged into HC&S in 1948, bringing the total fee-simple HC&S acreage to 51,137 acres, with 22,855 acres planted to cane. This permitted the combined plantation to harvest 11,119 acres in 1950. Over a 12-year period, the plantation had replaced variety H 109, principally with H 32–8560 (11,461 acres), H 37–1933 (7,040 acres), and H 38–2915 (5,446 acres). These varieties were, in turn, replaced by the late 1960s with H 50–7209 (15,522 acres), H 50–2036 (8,375 acres), and H 57–5174 (3,741 acres). Between 1950 and 1968, HC&S gradually increased its acreage harvested from 11,119 acres to 14,679 acres. Over the same period its sugar production increased from 142,482 tons to 198,536 tons, and at the beginning of 1969 it had 29,688 acres of cane under cultivation. The total labor force in 1969 was 1,437, down dramatically since 1950, when the plantation had 3,242 employees. As a result, plantation labor efficiency had more than tripled, from 43 tons of sugar per employee in 1950 to 138 tons of sugar per employee in 1968.[117]

Irrigation

In 1930 all of HC&S's cane land was irrigated, with 11 percent of the fields receiving only surface water and 89 percent with access to both surface and pump water. The average daily irrigation amount was about 200 million gallons. Three main irrigation ditches provided the surface (mountain) water: the Wailoa Ditch on the east, the 'Haikū Ditch through the middle of the plantation, and the Waihe'e Ditch on the west. The Waihe'e Ditch, owned by Wailuku Sugar Company, provided slightly more than half of its water to HC&S and the remainder to Wailuku Sugar Company. East Maui Irrigation Company, owned by HC&S and MAC, was responsible for development, management, and delivery of waters from the Wailoa and 'Haikū ditches to the plantations. HC&S had 25 miles of concrete-lined main ditch (11 feet top width, 4 feet bottom width, 5 feet deep); 26 miles of unlined main ditch; 1 mile of cut stone–lined main ditch (3 × 3 feet); 3 miles of concrete-plaster-lined straight ditch (5 feet top width, 3 feet bottom width, 4 feet deep); 7 miles of wooden flume ditch (2 × 2 feet); 11 miles of steel pipeline (30 inches in diameter); and 1 mile of wooden stave pipe (12 inches in diameter).[118]

From 1926 to 1930, mountain water from these three ditches provided an average of 121 mgd to HC&S, but average annual flows varied by more than a factor of two: 1926 (70 mgd), 1927 (131 mgd), 1928 (137 mgd), 1929 (110 mgd), 1930 (157 mgd). By 1930 HC&S had converted all its pumping stations from steam to electric or diesel power. Its one diesel and nine electric pumping stations had a total pumping capacity of about 142 mgd.[119]

In 1939 average HC&S irrigation amounts were similar to those in 1930, about 200 mgd. The five-year average delivery of mountain water from the three main ditches was 119.5 mgd, almost equal to the 121 mgd five-year average in the late 1920s. Its pumping stations could deliver up to an additional 178 mgd, a substantial increase from the 142 mgd in 1931, primarily the result of installing two new pumping stations with a total capacity of 36 mgd. This provided an "ample factor of safety in event ditch supply should drop unduly," and at times the pumps had "been called on to deliver as much as 52%" of the plantation's irrigation water.[120]

The main components of the plantation water conveyance system had changed little since 1931. However, because of the corrosive effects of sand in the fast-moving water of the concrete-lined irrigation ditches, substantial efforts were being made to remove the sand with settling basins. The plantation also made significant strides in increasing the efficiency of irrigators and the effectiveness of the irrigation water they applied. Beginning in the mid-1930s plant fields were changed to more efficient long-line irrigation, and by 1938, about 70 percent of the field area was in long-line and border irrigation systems, leaving only 30 percent in the old contour system. The plantation still applied 27–30 irrigations to a full crop, with an average interval of 17 days, but irrigator productivity had increased from about 10 to 30 acres per day.[121]

As of 1951 the plantation irrigation system, which now included both the HC&S and MAC systems, had a capacity of 448 mgd of mountain ditch water. The combined plantation had 18 pumping stations with 34 pumps, including eight booster pumps to raise water to higher levels of the plantation. The total pumping capacity was 230 mgd with a booster pump capacity of 80 million gallons daily. Electric power for the pumps came from the power plants of the two plantation mills at Puʻunēnē and Pāʻia, as well as from two hydropower plants.[122]

Three types of furrow irrigation were in use at this time. The level ditch system was used in fields with slopes of 2 percent or less. In fields with slopes over 2 percent, the herringbone system used concrete flumes, and the continuous long-line system used aluminum flumes. The plantation could apply approximately 6 inches of water per irrigation in about 40 rounds per crop (240 inches per crop), a substantial increase over the 27–30 irrigations per crop in the late 1930s. Tensiometers were used to control irrigation intervals before the ripening phase, and leaf sheath moisture determined by crop logging was used during ripening prior to harvest. Aerial photography was used to locate dry spots in the fields and to survey storm damage.[123]

From 1951 to 1968 the conveyance of mountain water by the 70 miles of ditches and tunnels in the East and West Maui systems increased from 448 mgd to 553 mgd. Ultimately the plantation's 16 pumping stations (with

The crop logging method developed by H. F. Clements of the University of Hawaiʻi and used by HC&S to control fertilization and ripening decisions was described as follows.

Starting when a plant or ratoon field has developed six leaves, which occurs within two and three and a half months, representative samples of leaves and sheaths are taken at thirty-five day intervals throughout the growing period of the crop. These samples are taken to the laboratory where they are prepared and analyzed for moisture, total sugar, nitrogen, phosphorus and potassium.

The results of these analyses, together with weather data, are posted on the crop log sheet which is kept for each field. The interpretation of this information determines the kind and amount of fertilizer the field requires and whether sufficient irrigation has been applied.

Seven months prior to the tentative harvest date, each field is put on a 14 day sampling schedule and the results of the moisture determinations are posted on a specially prepared chart called the ripening log. The objective of ripening control is to gradually reduce the percentage moisture in the leaf sheath to about 73% at harvest by withholding irrigation.[124]

41 pumping units) had a capacity of 272 mgd, with a booster pump capacity of 121 mgd.[125]

In 1968 the level ditch system was used for fields with slopes between 0.5 and 2 percent (15,100 acres). The herringbone system with concrete flumes was used on fields with slopes greater than 4 percent (12,549 acres). Less-important systems included the level ditch system for fields with slopes of 2 to 4 percent (1,365 acres), and the continuous long-line system (308 acres). Overhead sprinkler systems were used on 400 acres. Irrigation amounts were normally about 6 inches per round, and an average of 42 rounds were applied per crop—very similar to the irrigation applied in 1951. This was considerably more irrigation water than the 140–180 inches the MASI and lysimeter experiments on HC&S estimated that a two-year crop actually used. The additional irrigation was needed because of unavoidable inefficiencies in furrow irrigation systems, but most of the excess water percolated below the root zone to help recharge the basal aquifer.[126]

Tillage and Weed Control

In the late 1920s and early 1930s HC&S normally ratooned a field several times until its yield declined sufficiently to warrant a new planting, on average every 10 years. Land preparation for new plant fields usually consisted of plowing 18 inches deep on the contour with five Fowler steam plows and four 60-horsepower diesel Caterpillar tractors. Water carts carrying 800 gallons and pulled by 30-horsepower Caterpillar tractors were used to service the steam plows. Most other cultivation was with hoes. Caterpillar tractors were also used for miscellaneous operations like hilling up ratoons, clearing land, maintaining roads, building reservoirs, and subsoiling, although subsoiling was normally done with animal traction.[127]

By 1939 the plantation had a fleet of five 70- to 75-horsepower diesel and four 60-horsepower gasoline crawler tractors for heavy plowing, replacing its now antiquated steam tractors with their cable plows, water carts, and large labor requirements. It also had acquired nine smaller (10- to 40-horsepower) tractors for harrowing, light plowing, and ditching. Ten mule-drawn weeders with discs that passed on each side of the line of cane were still in use in 1939. The discs removed the weeds between the rows of cane; with repeated passes, the weeder threw soil away from the cane row, maintaining the furrow to conduct irrigation water. One or two hoeings were then sufficient to suppress weed growth until the cane canopy closed in, shading out any further weeds that might germinate.[128]

By the early 1950s heavier equipment allowed land to be plowed more deeply in preparation for planting. Plant fields were subsoiled to 24 inches, deep plowed to 22 inches with 42-inch disc plows, and harrowed. All these operations used 130-horsepower diesel tractors operating in three shifts. Two 75-horsepower diesel tractors with bins for seed cane and fertilizer opened furrows with moldboard plows, planted, applied initial fertilizer, and covered the seed in one pass. The seed cane was cut by the planter to 22-inch seed pieces and sprayed with phenyl mercuric acetate as it slid down the chute into the furrow. This one-pass method of planting, fertilizing, treating with fungicide, and covering represented a tremendous labor saving compared with prewar methods.[129]

After they were harvested, ratoon cane fields were ripped, their irrigation furrows were reshaped, and anhydrous ammonia was applied, all in one pass with a 50-horsepower tractor. This allowed the first irrigation to be applied within 10 days of harvest. Replanting was conducted as soon as regrowth was sufficient to identify where stools had died. The same variety that was already in the field was used for replanting, except when replanting occurred in the winter, when variety H 44–3098 was used due to its rapid germination and vigorous early growth.[130]

By the early 1950s mechanical weed control had been abandoned, and knapsack sprayers were used to apply CADE diesel oil emulsion, a contact herbicide. The CADE emulsion was prepared at the plantation's central mixing station and was hauled to the field in tank trucks to fill the knapsack sprayers. A total of six or seven applications were normally made at three- to five-week intervals, beginning about three weeks after the first irrigation.[131]

In 1969 the plantation maintained a fleet of 83 crawler tractors for clearing, plowing, planting, harvesting, ratooning, and other heavy field operations. Thirteen wheel-type tractors were used for lighter farming and cultivation work. Land preparation still included subsoiling to 24 inches, but the plantation had replaced the 42-inch disc plow (used in the early 1950s) with two passes by heavy 5-tine plows, followed by a landplane to level the field for accurate surveying of rows. Planting and ratooning methods had not changed significantly since 1951; however, chemical weed control had evolved. Ametryne or Diuron was sprayed by airplane one or two days after the first irrigation in both plant and ratoon fields, and Ametryne was applied between the cane rows with backpack sprayers once or twice before the cane canopy closed in and shaded out weeds.[132]

Fertilization

Fertilizer practices in the 1930s were similar for plant and ratoon crops, beginning with 1,000 pounds per acre of mixed fertilizer (2 percent nitrogen, 19.75 percent phosphate, 18.5 percent potash) in the furrow for plant cane or on top of the stubble for ratoon cane. During the first season, 600 pounds per acre of ammonium sulfate (20.5 percent nitrogen) was applied in three doses of 200 pounds each, either in irrigation water or by hand, at two-month intervals. In late-planted or late-ratooned fields, the application rate was reduced to 400 pounds of ammonium sulfate. During the second season and as soon as weather permitted, sodium nitrate (15.5 percent nitrogen) was applied in two 350-pound doses at six-week intervals in irrigation water.[133]

Fertilizer experiments conducted in the 1930s, along with soil and water analyses, convinced HC&S managers that their soils had adequate potash levels, and no fertilizer potash was needed. By 1939, all fields initially received 35 pounds of nitrogen and 146 pounds of phosphate per acre as ammonium phosphate. This was supplemented by ammonium sulfate necessary to bring total nitrogen to 200–240 pounds per acre.[134]

By 1950, about two-thirds of the nitrogen fertilizer (about 150 pounds per acre) was normally applied in split applications during the first six months of the crop. Crop logging was used to determine the amount applied during the second season, normally about 12 months prior to the anticipated harvest date. Anhydrous ammonia was injected into the soil of ratoon crops by the tractor that reshaped the furrows. Ammonium sulfate was applied by hand for the second and third applications. Aqua ammonia was added via irrigation water for the second season fertilization; however, the plantation planned to replace all ammonium sulfate fertilizer with aqua ammonia as soon as sufficient supplies could be obtained from mainland sources.[135]

Phosphate and potash fertilizers were applied only when crop logging indicated that they were deficient. Normally, 100 pounds per acre of phosphate as raw rock phosphate was applied with the seed. Potash was required on only about 6 percent of the plantation's cane lands (applied by hand at a rate of 100 pounds per acre as soon as the crop had emerged). If crop logging indicated that additional potash was required, it was applied by hand or in irrigation water at a rate of 100 pounds per acre.[136]

By 1968 fertilizer rates were being determined by several factors, including crop logs, soil analysis, aerial and ground observations, and field history. A plantation average of 340 pounds of nitrogen per acre (a substantial increase over 1951 rates) was applied as aqua ammonia in irrigation water, normally in six doses between the first and the eleventh months of growth. Phosphorus was typically applied by airplane as triple super phosphate in amounts required to maintain extractable soil phosphorus above 85 pounds per ac-ft, as estimated from Experiment Station soil tests. Potash was applied in irrigation water at rates determined by crop logging and soil testing.[137]

Animals

In 1930 HC&S maintained about 700 horses and mules to assist with field work, including 210 saddle horses, 105 draft horses and mares, 130 draft mules, 60 pack animals, and 195 breeding stock. They were used for hauling seed cane, trash, and fertilizer in the fields; pulling oil wagons for steam plows; subsoiling ratoons; hauling cane and portable track during harvest; hauling firewood; and clearing land. The animals received three feedings per day of chopped cane tops and mixed feed composed of ground oats, kiawe bean meal, pineapple bran, bagasse, alfalfa meal, and molasses. The cattle ranch also had 865 head of beef and dairy cattle.[138]

At the end of the decade, however, the HC&S horse and mule herds had been reduced to about 420 head, reflecting the continuing emphasis on mechanization. The plantation dairy herd had been improved and consisted on 316 purebred Jersey and Holstein cows, of which 125 were milked daily, producing 1,100 quarts of milk for plantation families.[139] Later, the plantation herd of saddle horses was eliminated, and by 1951 only about 30 mules and horses remained for use as pack animals. The plantation continued to maintain a herd of about 1,400 Angus cattle on 8,000 acres of ranch land. It also continued to milk about 200 head of dairy cows each day, but the milk was sold commercially.[140]

Cane Harvest and Hauling

In 1930 all cane was hauled to the factory on the factory railroad, consisting of 77 miles of permanent track and 11 miles of portable track, eight oil-fired steam locomotives, and 1,092 cane cars. Four gasoline-driven crawler cranes were used to pick up bundles of hand-harvested cane and place them onto the cane cars. In the off-season, these cranes were used for excavation and lining irrigation ditches.[141] Within 10 years the plantation had increased its permanent track to 83 miles, reduced its portable track to eight miles, replaced three of its steam locomotives with two diesel-electric locomotives, and reduced its fleet of cane cars to 865.[142]

The 1938–1939 crop marked the beginning of mechanical cane harvesting at HC&S. The previous method of cutting by hand and loading with cranes continued in one harvesting gang, but 43 percent of the crop was harvested mechanically. The cane was cut and dragged into windrows about 60 feet apart with a rake pulled back and forth across the field by two diesel tractors. Portable track was then laid parallel to the windrows, and the cane was loaded onto cane cars with cranes.[143]

During 1950 and 1951 the plantation railroad system was closed down, and cane was hauled to the Puʻunēnē and Pāʻia factories with huge Tournahaulers (refer back to Figure 4.11). This revolutionary change led to dramatic cuts in both the cost and the labor needed to move temporary track and maintain the plantation rail system. Thirteen Tournahauler units, each with 40- or 50-ton capacity trailers, had replaced the 865 cane cars in use in 1939. But major improvements in the plantation road system were required to handle the heavy Tournahaulers and allow them to reach every field.[144] Harvesting was conducted in three eight-hour shifts by two field gangs, one for the Puʻunēnē and another for the Pāʻia factory. All cane fields were burned approximately 24 hours before harvest. After the cane had burned, 80-horsepower bulldozer push rakes severed the cane at ground level and pushed it into windrows for loading (see Figure 4.12). As soon as the bulldozer rakes had formed a windrow, the cane furrows alongside the windrow were leveled to allow Tournahaulers to maneuver alongside it. The cane in the windrows was lifted into the trailers by crawler cranes with grabs. While the cranes were loading one windrow, the bulldozer rakes were forming the next windrow nearby.[145]

Harvesting methods in the late 1960s were similar to those in the early 1950s, except that six 230- to 270-horsepower crawler tractors equipped with push rakes were used to cut the cane and push it into windrows, replacing the 80-horsepower tractors. In addition, "brooming" or *liliko* rakes—50- to 120-horsepower tractors with bulldozer rakes on the front and spring tooth rakes on the back—were used to gather the cane missed by the heavier equipment.[146]

In 1969 the plantation had 18 cane hauling rigs, each with a truck tractor and a 40-ton semitrailer. The network of cane roads allowed cane from any field to be transported to either the Pāʻia or Puʻunēnē factory. These roads, as well as in-field roads, were maintained by six motor graders, a road roller, and two semitrailer road watering trucks.[147]

Factory Operations

In 1930 the HC&S factory at Puʻunēnē had motor-driven car unloaders and rakes to feed its mill. Each of its two tandem sets of milling equipment included 12-roll mills (four sets of three mills) preceded by revolving knives, a 2-roll crusher, and a shredder. The capacity was 70 tons of cane per hour for each tandem, though the normal rate was about 58 tons per hour. The steam plant consisted of one large boiler operating at 450 psi and 16 smaller boilers at 150 psi. The factory electric plant had two 750-kilowatt generators. The clarification process used five primary and two secondary juice heaters, and two primary and two secondary clarifiers. No filter presses were used. Instead, 14 centrifugal mud separators removed mud from material that settled in the clarifiers. Two quadruple-effect evaporators were used to concentrate the clarified juice. Eight vacuum pans, each served by an individual steam condensation system and vacuum pump, were used to concentrate the juice. Sugar was then passed to 32 crystallizers for crystal growth. Eleven centrifuges driven by electric motors were used to produce high-grade (No. 1) sugar, and low-grade sugar was produced by 40 water-driven centrifuges. Sugar and molasses storage facilities could accommodate 14,000 tons of sugar and 2,000 tons of molasses.[148]

By 1939 some important improvements were made in the factory at Puʻunēnē. In an effort to remove soil, rocks, and other foreign material from the mechanically harvested cane, a three-stage cane cleaning system had been installed on one of the factory's two tandem mills. It consisted of "a revolving drum where dirt and rocks [were] shaken out, thence over a 30 [foot] long by 10 [foot] wide table where much sand and small rock [was] eliminated." The cane was then conveyed "over a 40 foot washing table where water from the pan and evaporator condensers [was] applied to wash off the remaining soil."[149] In 1931 the factory replaced its 14 centrifugal mud separators with three revolving strainers to remove solids—called *cush-cush*—from the juice, and three 8 foot × 12 foot filter presses.[150]

By 1951, when the loaded Tournahaulers, each carrying 25 to 32 tons of cane, arrived at the Puʻunēnē and Pāʻia mills, a crane unloaded the vehicles in a single operation. "This [was] done by engaging a hook on the crane under metal beam called a "strongback" that [ran] the entire length of the trailer and [was] fastened to a chain net . . . slung under the cane and [was] firmly attached to the other side of the trailer body. When the crane [was] put into operation the cane [was] rolled out of the trailer and over the dumping wall."

By 1950, because of their implementation of mechanical harvesting, both the Puʻunēnē and Pāʻia mills had installed cane washing systems for all their cane. At the Puʻunēnē factory the cane was fed into the cane cleaner consisting of "a carding drum . . . to eliminate large bunches," and a "3 [foot] wide hydraulic gap . . . to allow rocks to sink through and be carried away"; the cane was then "dropped on a splitter" that could direct the cane to either or both of the two tandem mills. Each tandem had its own system to shake the stools and strip the trash from the cane. The Pāʻia factory had a different cane cleaning system that passed the cane over rollers to remove soil then through a "hydraulic gap" across which the cane floated while the rocks sank.[151]

In 1950 the Puʻunēnē and Pāʻia mills had capacities of 3,300 and 2,400 tons of cane per 24 hours, respectively. Both factories had double tandem mills preceded by revolving knives to cut the cane into small pieces and a crusher to produce a uniform blanket of cane entering the mills. Both factories used rotary vacuum filter presses to process settlings from the clarifiers. The Puʻunēnē factory used two quadruple effect evaporators and eight vacuum pans, while Pāʻia used a single quintuple evaporator and five vacuum pans. The Puʻunēnē factory had warehouse capacity of 14,000 tons of bulk sugar while the Pāʻia factory shipped its bulk sugar to the port of Kahului and stored it there.[152]

In 1969 the capacities of the Puʻunēnē and Pāʻia mills had been increased to 6,200 and 3,700 tons per 24 hours, respectively. The Puʻunēnē cane cleaning plant and the tandem mills at both plants were similar to their 1951 configurations. However, a new and improved Pāʻia cane cleaning plant, similar to the one at Puʻunēnē, was installed in 1962. Other factory improvements between 1951 and 1968 included more powerful cane shredders and new juice clarifiers, rotary vacuum filter presses, and centrifugals at both factories.[153]

In 1969 the plantation still operated its rotary lime kiln, producing hydrated lime for the two plantation mills and selling the excess to other sugar companies.[154]

Electricity Generation

By 1931 HC&S had two steam power plants and one hydroelectric plant with a total capacity of 12,000 kilowatts (KW) if operated at 80 percent power. The majority of the plantation's power was generated at the plantation's central power plant. In addition, a diesel-driven generator was available for use during periods of great irrigation demand during the dry season.

Another smaller electric power plant was located at the sugar factory and supplied the power needed to manufacture sugar. In addition to the central power plant and the factory power plant, the Kāheka hydroelectric power plant was located seven miles away. It was driven by three turbines that received water from the Wailoa Ditch via a three-mile-long riveted steel pipe with 660 feet of head. This water was then discharged into the Lowrie Ditch.

In the early 1930s electric power was consumed by irrigation pumps (6,000 KW), the sugar factory (1,500 KW), and the general lighting and power requirements (400 KW) of the plantation's laborers' quarters, livestock feed mill, dairy, carpenter shop, refrigerator plants, garage, machine shops, movie house, and so forth. About 900 KW of excess power was sold to Kahului Railroad Company and the Maui Electric Company.[155] By 1939 HC&S was using a new "800 [horsepower] water tube boiler, to supply steam at 300 pounds pressure to the new mill power house turbine." This turbine powered a new 4,000 KW generator, which was the key element in increasing generation capacity to 16,000 KW.[156]

After the 1948 merger of the HC&S and MAC, the combined plantation had five power plants: the central steam power plant, the steam power plants at the Puʻunēnē and Pāʻia factories, and two hydropower plants. In 1950 the Puʻunēnē factory had 5,500 KW generating capacity, the Pāʻia factory had 6,000 KW, and the two hydropower plants had a total of 6,400 KW. The central power plant near the Puʻunēnē factory was now used as a "stand-by unit," with 6,250 KW capacity.[157]

By the late 1960s the combined plantation had just four power plants; the central power plant near the Puʻunēnē factory had been eliminated, and all power was generated by the Puʻunēnē and Pāʻia factory steam plants and two hydropower plants. The Puʻunēnē factory plant contained a 4,000 KW and a 7,500 KW steam plant. The Pāʻia plant had two 4,000 KW steam plants. Each one was supplied by steam generated from bagasse during the harvest season and fuel oil at other times. The two hydropower plants used water from the Wailoa Ditch. The Kāheka plant used 85 mgd with a drop of 660 feet to the Lowrie Ditch to produce 4,000 KW. The Pāʻia hydroplant generated 800 KW with 45 million gallons a day that dropped 293 feet to the Kauhikoa Ditch.[158]

Labor

In 1931 about 7,000 men, women, and children lived in 1,685 houses at 26 camps. Of the houses, 75 were for skilled employees, 400 for semiskilled, and 1,210 for laborers. All the camps had electricity. Two-thirds had sewer systems with septic tanks, and all homes had "filtered or pure mountain water." All laborers earning $100 or less per month had free medical and hospital treatment, nurses visited the camps regularly, and a hospital was maintained on the plantation. Four public schools, three Japanese language schools, 12 day nurseries, and 10 churches were provided by the company. A dairy and a cattle ranch supplied milk and beef to workers at cost. Recreational facilities included a "swimming tank," gymnasium, three theaters (for "talkies"), and a number of baseball and athletic fields.[159]

The total census for the HC&S Plantation in 1939 was 7,973, of which 5,541 of the workers were U.S. citizens. The largest ethnic group was Japanese, with 4,173 individuals (more than half of them children). The next largest ethnic group was Filipino, with a total of 1,930; the third largest was Portuguese, with 1,009.[160] The workforce lived in 30 camps with 1,545 houses distributed across the plantation. At the end of the decade the plantation was delivering about 575 cords of firewood per month to the camps for cooking and heating water, but it had begun to supply 14,000 gallons of kerosene per month for cooking and 475 gallons of diesel per month for heating water, and it predicted that this trend would continue until no more wood would be used in the camps. To assure adequate nutrition for the single men in the workforce, in 1938 a cafeteria—"modern in every respect"—was constructed as "a non-profit project." As in 1931, the plantation maintained a "modern and complete central hospital with six resident nurses and a dietician working under a staff of three physicians." Two nurses and a dietician also provided baby clinics and attended to "minor ills of the laborers and their families."[161]

During World War II military service and defense industries took plantation labor and caused severe labor shortages; high school students were required to work as field laborers on Fridays and Saturdays. The military imposed a total blackout on the island and took 3,800 acres of HC&S land for construction of Puʻunēnē and Kahului Naval Air Stations. In addition, the company sold 45 percent of the electricity it produced to the military and set up Victory Gardens near its villages.[162]

After the war ended, major changes began to occur in labor relations. The International Longshore and Warehouse Union (ILWU) organized sugar workers throughout the territory, and the last large group of Filipino immigrants arrived under the auspices of HSPA. Soon thereafter, it became clear that economies of scale could be achieved by merging the neighbor plantations HC&S and Maui Agricultural Company. The merger was completed in 1948. Frank Baldwin was named president, his son Asa was manager, and Alexander & Baldwin owned approximately 34 percent of the shares in the new company.[163] Also in 1948, Frank and Asa Baldwin, in part responding to the new labor relations in the industry, announced a long-term plan to create a "Dream City" on unused HC&S land in Kahului, allowing company employees to purchase their homes and gradually eliminating the company's 66 plantation villages and 4,000 homes. The intent of the project, which was sponsored by HC&S and Kahului Railroad (KRR), was not to make a profit, but to eliminate the paternalistic landlord-tenant relationship with company employees. Kahului Development Company, Inc. (later A&B Properties, Ltd.) was formed to manage the project, which worked quickly, with the first families moving into their homes in Kahului in 1950. By 1985 a total of 3,000 lots had been sold, of which more than 2,000 were purchased by employees of HC&S and other A&B companies. Additional changes occurring in the postwar years included the merger of Pāʻia and Puʻunēnē company hospitals in a remodeled facility in Puʻunēnē and the merger of the HC&S Plantation store system with the KRR merchandise department, forming A&B Commercial Company.[164]

In 1951 the combined plantation had 3,242 employees, including about 520 supervisory, medical, and technical employees. Its population of 3,843 men, 2,726 women, and 4,176 children lived in 2,445 houses "rented at low rates and . . . fully equipped with electricity and running water." The largest three ethnic groups living in plantation housing were Japanese (5,271), Filipino (2,273), and Portuguese (1,370). The plantation hospital served an average of 70 patients per day. In 1949 it had added "an air-conditioned wing with two modern operating rooms, [a] maternity ward and kitchen." Five elementary schools and three high schools (one Catholic and two public) were available to plantation families. "Two community associations controlled by labor, management and community representation sponsor[ed] year-round athletic and recreational programs."[165]

From 1951 to 1953 HC&S held several large celebrations to commemorate the fiftieth anniversary of the Puʻunēnē factory. In 1956 HC&S closed the Puʻunēnē Plantation hospital, with the new Central Maui Memorial Hospital taking over care of plantation personnel, and in the early 1960s it turned over its parks and playgrounds to Maui County. However, labor relations deteriorated with the industry-wide strike in 1958, which lasted 126 days and caused HC&S to lose money in that year.[166]

The death of Frank Baldwin and retirement of HC&S manager Asa Baldwin in 1960 marked the end of the Baldwin family's leadership of the company. From 1961 to 1967 HC&S, in cooperation with Joseph E. Seagram & Sons, operated a rum distillery at the Puʻunēnē factory. In 1966 the KRR trains that had long transported raw sugar and molasses to the port of Kahului were replaced by trucks.[167]

At the beginning of 1969 HC&S had 1,174 hourly employees and 263 supervisory, technical, clerical, and medical staff. This represented a 56 percent reduction in employees since 1951. The population of the plantation's camps or villages had declined steadily from the mid-1950s numbers, and many employees had moved to the company-sponsored, 2,000-home New Kahului development and elsewhere in the area. The number of plantation houses had decreased from 2,445 in 1951 to only 720 units in 1968.[168] All employees and their families were covered by a company medical plan with a minimal cost to employees (12 percent of total cost), as well as a free company dental plan for hourly employees' children. Plantation schools had also been eliminated, and children attended both public and parochial schools. Public and private community and junior colleges were also available nearby. Two member-controlled community associations coordinated recreational and athletic programs available to plantation employees.[169]

Summary

The decade of the 1930s began on a positive note, with the sugarcane industry producing its first million-ton sugar yield. However, low sugar prices during the Great Depression, the passage of the Jones-Costigan Act (Sugar Act), and labor and materials shortages during World War II created continual pressure for the industry to reduce its labor force and increase efficiencies. Though cane was still cut by hand, mechanical loading onto railcars gained popularity in the 1930s. Hand harvesting was replaced by cable-drawn Waialua rakes in the 1930s and push rakes mounted on bulldozers in the 1940s and 1950s. In the 1950s plantation railroads were replaced by large rubber-tired cane haulers. The booming U.S. economy of the 1950s and 1960s buoyed Hawaiʻi's sugar companies. Sugar Act provisions prevented wild swings in sugar price and stabilized the returns. This allowed the Experiment Station of the HSPA to make unprecedented progress in sugarcane breeding, soil preparation, disease and insect control, chemical weed control, improvement of furrow irrigation systems, soil fertility and plant nutrition, mechanized harvesting and transportation, and factory operations. These technological advances allowed plantation owners to reduce their workforces and, in combination with steadily increasing yields per acre, greatly increased the amount of sugar produced per employee. Unsatisfied with their shrinking share of profits, labor demanded increased wages, and major strikes in 1945–1946 and 1958 reduced harvests and resulted in industry-wide wage concessions.

On Maui, in 1930 over 7,000 men, women, and children lived on HC&S land in more than 25 camps. Cane was still cut by hand and hauled to the mills on plantation railroads. The plantation maintained large herds of horses and mules for field work, and it produced beef, pork, and dairy products for its employees. The HC&S factory at Puʻunēnē produced more sugar per year than any other in the Hawaiian Islands. Mechanical loading of railcars began in the late 1930s, and mechanical harvesting began in the 1940s and 1950s, requiring factories to make major investments in machinery to separate rocks and soil from the cane and remove impurities from the juice. Large cane haulers called Tournahaulers replaced railroads in the early 1950s.

HC&S merged with Maui Agricultural Company in 1948 to gain economies of scale in field operations and management. However, economic pressures forced continued reduction in plantation labor forces throughout the 1950s and 1960s, and the workforce of the combined plantation declined from over 3,200 in 1951 to less than 1,500 in 1969. Continuing to streamline operations, HC&S gradually reduced its stock of employee housing, encouraging employees to purchase houses on favorable terms in the New Kahului development.

Major changes in field operations also occurred during this period. Irrigation expanded as the plantation increased the capacity of its groundwater pumping stations. Furrow irrigation became more efficient as level ditch, continuous long-line, and herringbone systems were introduced. Tensiometers and pan evaporation were used to guide the frequency of irrigation. Never content with the high labor cost of irrigation, HC&S experimented with several semiautomatic and automatic irrigation systems as well as sprinkler irrigation. Fertilizer rates increased with the post–World War II rise in the availability of various fertilizers, and soil and plant analyses were used to manage fertilizer application and ripening. A variety of herbicides replaced cultivation and hand hoeing for weed control in the 1950s. Airplanes were used to apply fertilizers and herbicides. All these technological advances kept HC&S profitable throughout most of the 1950s and 1960s, but more difficult times lay ahead.

5

DRIP IRRIGATION AND NEW DISEASE RESISTANT VARIETIES SAVE HC&S—1970 TO 2014

Since the 1970s the Hawai'i sugarcane industry has faced a number of challenges. The most fundamental issue: the price of raw sugar on the domestic and international markets did not keep pace with the industry's costs for labor and other inputs. As a result, sugar growers were forced to cut costs wherever possible in order to keep the cost of production below the going price for raw sugar and the by-products of sugarcane. The industry needed to reduce its labor costs while retaining its highly skilled workers, but that became increasingly difficult in the face of competition from tourism, government, and other employers. Fortunately, conversion from furrow to drip irrigation substantially reduced the labor needed to manage irrigation, fertilization, and tillage. This single far-reaching change in sugarcane production systems allowed most of the state's irrigated plantations to survive into the 1990s.

Environmental issues also became a greater concern for the industry. Public pressure increased, especially on Maui, to reduce the smoke from mills and pre-harvest burning of cane fields. Demands also increased for the industry to share its water resources with the islands' rapidly growing population, diversified agricultural producers, and the environment—including reducing withdrawals from streams in order to maintain environmental flows.

Sugar Prices and the Decline of the Hawai'i Sugar Industry

Much of the stability in sugar prices from the 1930s to the early 1970s was due to the sugar provisions of the Jones-Costigan Act (Sugar Act) of 1934 and its 1937 and 1948 amendments. In compliance with the act, each year the government determined how much sugar the nation would need and allocated quotas to domestic and foreign sources. It then levied an excise tax on refiners and distributed the proceeds among domestic producers as compensation for respecting production quotas, assuring equitable treatment of beet and cane sugar producers. As the result of a 1936 Supreme Court decision, Congress amended the Sugar Act, separating the excise tax from payments to domestic producers. This version of the law, which took effect in 1937, was extended several times until it was replaced by the Sugar Act of 1948, which changed quota allocations and gave preference to Cuban sugar in return for its cooperation in supplying the United States with additional sugar during World War II. The 1948 act was extended to 1960, when imports from Cuba were suspended as a result of the Cuban revolution. This act remained in effect until the passage of the Sugar Act of 1974.[1]

The Sugar Act maintained sugar prices at levels profitable to most of the Hawai'i industry from the late 1930s until the mid-1970s. However, world sugar supplies fell below world demand in the early 1970s, causing dramatic price increases. World raw sugar prices averaged a record 57 cents per pound in November 1974, leading Congress to eliminate import restrictions and excise taxes on imported refined sugar, as well as remove domestic sugar quotas and acreage allotments. In response, sugar production increased worldwide, causing average annual raw sugar prices to decline to less than 12 cents per pound in 1976, and remain at less than 10 cents per pound in 1977–1979. In an effort to help the beet and cane sugar industries, an interim government price support program was implemented in 1977, and the Food and Agriculture Act of 1977 established raw sugar loan rates of 13.5 and 14.73 cents per pound for the 1977 and 1978 crops, respectively. If markets fell below those loan rates, processors could default on their loans, forfeiting their sugar to the Commodity Credit Corporation (CCC). In order to maintain prices near or above the loan rates and minimize forfeitures, import duties and fees were levied on imported sugar. A similar program was implemented for the 1979 crop.[2]

Because sugar prices from 1976 through 1979 were simply too low to cover the costs of production, a number of the Hawai'i industry's less competitive companies were unable to survive. Between 1970 and 1980, 13 of the industry's 27 companies closed. The industry on the island of Hawai'i was hard hit, as Pā'auhau Plantation Company (1972), Hāmākua Mill Company (1974), Kohala Sugar Company (1975), Honoka'a Sugar Company (1979), and Laupāhoehoe Sugar Company (1979) all closed their doors. The hardship was lessened somewhat when the Davies Hamakua Sugar Company was formed, taking over most of the lands taken out of production on the Hāmākua coast. While the lands of the Kohala Sugar Company were permanently taken out of sugar production, one mill located at Honoka'a processed all the cane formerly processed in four mills, and the cane produced closer to Hilo was processed by the Hilo Coast Processing Company (HCPC), a cooperative made up of Mauna Kea Sugar Company and some independent growers. On Kaua'i, Kīlauea Sugar Company closed in 1971 and Grove Farm Plantation ceased operations in 1974. But Lihu'e Sugar Plantation and McBryde Sugar Company leased parts of the Grove Farm acreage, reducing the impact on the state's sugar production. On O'ahu, 'Ewa Plantation Company and Kahuku Plantation Company ceased operations in 1970 and 1971, respectively. O'ahu Sugar Company incorporated the 'Ewa sugar lands, but the Kahuku lands were permanently taken out of production. Maui's three plantations, HC&S, Wailuku Sugar Company, and Pioneer Mill Company, remained in operation throughout the 1970s.[3]

Because of unstable world production rates, the average annual world price of raw sugar rocketed back to 29 cents per pound in 1980. U.S. Congress, bowing to pressures from consumer groups and the processed food industry, canceled the price support program for the 1980 and 1981 crops, assuming that the domestic industry would flourish with such high prices. The U.S. sugar industry knew from previous booms and busts that it could not live with the continuing price uncertainty. In a series of high-level meetings, representatives of the U.S. cane, beet, and corn sweetener industries concluded that the U.S. sweetener industry needed to be part of the Farm Bill that supported other major crops. They also realized that Congress would not allow direct subsidies to the industry, and the program would have to operate at no cost to the government. The 1981 Farm Bill reestablished raw sugar loan rates, which rose from 17 cents per pound in 1982 to 18 cents per pound in 1985. In addition, import duties were increased and a system of country-by-country import quotas was established. The goal of Congress was to protect the U.S. sugar industry from fluctuations in world sugar prices by maintaining market prices near a "market stabilization price" above the loan rate to discourage default sales to the CCC. Partly because of the import quotas in the Farm Bill, the world price of raw sugar plunged to an average annual 8.4 cents per pound in 1982 and to a low of about 4 cents per pound in 1985—far below the U.S. loan and market stabilization prices. The 1985 Farm Bill continued most of the provisions of the 1981 act with loan rates of no less than 18 cents per pound though 1990. However, sugar refiners and the processed food industry complained that, by paying the domestic price of more than 18 cents per pound, they and domestic consumers were subsidizing domestic sugar producers. U.S. Department of Agri-

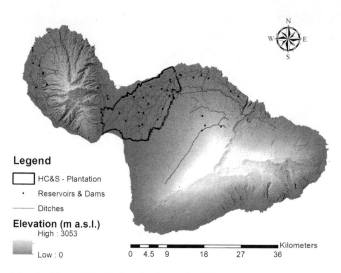

Legend

☐ HC&S - Plantation

· Reservoirs & Dams

——— Ditches

Elevation (m a.s.l.)

High : 3053

Low : 0

0 4.5 9 18 27 36 Kilometers

Map 1. Map of Maui showing irrigation ditches, reservoirs, and HC&S land after incorporation of Wailuku fields.

culture (USDA) officials, aware of political realities, warned the industry that although the domestic sugar price support program was "no net cost" to the government, it could not be maintained in the long term. Higher costs to domestic food processing industries and the consumer, as well as international pressures to increase U.S. sugar imports to stabilize the economies of developing countries, were creating strong pressures on Congress.[4]

Responding to the volatile price support situation in the early 1980s, the industry redoubled its efforts to increase efficiency. During the first half of the 1980s the Hawai'i industry increased its per-acre sugar yields by 13 percent, reduced the direct cost of producing a ton of sugar by 15 percent, and reduced the number of worker-days needed to produce a ton of sugar by 20 percent. Nevertheless, sugar companies on the island of Hawai'i became increasingly uneconomical, with Puna Sugar Company (1982), Davies Hāmākua Sugar Company (1984), and Ka'ū Sugar Company (1986) ceasing operations. On Maui, Wailuku Sugar Company went out of business in 1988, but HC&S took over some of its cane lands along with its irrigation allocations from West Maui Irrigation, reducing the impact on Maui's sugar production (see Map 1).[5]

Another challenge for the domestic sugarcane and beet sugar industries was increased competition from other sweeteners. Back in 1988, high fructose corn syrup accounted for 31 percent of U.S. sweetener consumption.

Dextrose and glucose corn syrups provided 15 percent, and saccharine and other noncaloric sweeteners accounted for 12 percent of consumption, leaving sucrose with less than 50 percent of the U.S. sweetener market.[6]

Environmental challenges faced by the Hawai'i industry included public concerns about smoke from sugar mills and cane burning, use of crop protection chemicals on sugarcane fields, and water pollution. Other pressures included increasing costs of liability and workers' compensation insurance, urbanization, and the difficulty of hiring and retaining skilled employees. Public concerns about air quality and employee recruitment and retention were of special concern at HC&S. Production costs increased, and despite good per-acre yields, the industry as a whole lost money during late 1980s, suffering pre-tax losses of $10.3 million in 1988 and $5.4 million in 1989. These losses caused companies to cut corners and reduce costly inputs so that by 1991 sugar yields per acre had begun to decline across most of the industry. Some blamed the trend on varietal yield decline; however, analysis of plantation records revealed that the principal reason was harvesting cane too early, with other contributing factors including poor weed control and timing of operations, insufficient irrigation, and unusually wet weather on the Hilo-Hāmākua coast of Hawai'i.[7]

The 1990 Farm Bill maintained the loan rate at 18 cents per pound and set the market stabilization price at 21.95 cents per pound. In order to further stabilize imports, an import quota of 1.25 million tons of raw sugar and a two-tier tariff system were established. The 1996 and 2002 Farm Bills maintained the 18 cent per pound loan rate but raised the import quota and used marketing quotas to limit the amount that processors could sell domestically. This helped maintain domestic sugar prices above the minimum price support level, preventing the Commodity Credit Corporation (CCC) from having to purchase domestic sugar, thereby allowing the program to operate at "no cost" to the government. However, because of the North American Free Trade Agreement (NAFTA), since 2008 Mexican sugar has not faced quotas or duties. To address the potential for a U.S. sugar surplus caused by this and other trade agreements, the 2008 Farm Bill mandated a sugar-for-ethanol program requiring the USDA to purchase as much U.S. produced sugar as necessary to maintain market prices above support levels, to be sold to bioenergy producers for processing into ethanol.[8]

As a result of Farm Bill sugar programs, since the 1970s (with the exception of 1973–1976, 1979–1981, and 2009–2012, when world prices soared), prices received by U.S. sugar producers remained quite stable and substantially above world prices. U.S. prices in 1982–2009 were consistently between 18 and 24 cents per pound, while world prices fluctuated from about 4 cents per pound in 1985 to between 7.5 and 9.1 cents per pound in 2000–2004—less than half the average annual loan price for U.S. sugar during those years. Average annual world prices rose again to between 11.3 and 13.8 cents per pound in 2005–2009, still well below U.S. prices. But beginning in mid-2009 world raw sugar prices leaped. This led to average quarterly New York raw sugar prices greater than 30 cents per pound from late 2009 through mid-2012.[9]

Despite relatively stable domestic prices, Hawai'i's sugar companies continued to close in the 1990s. O'ahu Sugar Company ceased operations in 1994, and Waialua Agricultural Company closed in 1996—these were the last two remaining plantations on O'ahu. Davies Hāmākua Sugar Company and Mauna Kea Sugar Company closed in 1994, and Ka'ū Agribusiness Company halted its sugar operations in 1996, ending over 150 years of sugar production on the island of Hawai'i. On Kaua'i, Gay & Robinson purchased Olokele Sugar Company in 1994, and McBryde Sugar Company closed two years later. On Maui, Pioneer Mill Company closed in 1999, leaving HC&S as the only remaining Maui plantation.[10]

All told, at the end of the 1990s only four plantations remained. On Kaua'i, Gay & Robinson, Lihu'e Sugar Plantation (known as Amfac Sugar/East), and Kekaha Sugar Company (Amfac Sugar/West) continued to produce sugar through the 1990s. But the two Amfac plantations closed in 2000, and the last cane was harvested on Gay & Robinson in 2010, leaving only one operating sugar plantation in Hawai'i: HC&S.[11]

Experiment Station, Hawaiian Sugar Planters' Association (HSPA)

Funded by voluntary contributions from the Hawaiian Islands' sugar companies, the Experiment Station of the HSPA—like the plantations—was buffeted by sugar price fluctuations during the 1970s. World sugar prices shot up in 1974, leading to optimism in the industry, but this turned to pessimism as prices plummeted in 1975 and 1976. In response, the sugar agencies cut their contributions to the Experiment Station, which was forced to eliminate some 35 positions and scale back its research program.[12] Nevertheless, priority research continued. Smut, a major disease of sugarcane worldwide, was found in an Experiment Station Makiki field station in 1971. This potentially devastating disease could reduce yields of susceptible varieties

Plantations ceasing production between 1960 and 2010

Kaua'i
Waimea Sugar Company—1969
Kīlauea Sugar Company—1971
Grove Farm Plantation—1974
Lihu'e Sugar Plantation
  (Amfac Sugar/East)—2000
Gay & Robinson—2010

O'ahu
'Ewa Plantation
  Company—1970
Kahuku Plantation
  Company—1971
O'ahu Sugar Company—1994
Waialua Agricultural
  Company—1996

Maui
Wailuku Sugar Company—1988
Pioneer Mill Company—1999

Hawai'i
'Ōla'a Sugar Company—1960
Hilo Sugar Company—1965
Pā'auhau Plantation
  Company—1972
Olokele Sugar Company—1994

McBryde Sugar
  Company—1996
Kekaha Sugar Company (Amfac
  Sugar/West)—2000
Hutchinson Sugar
  Company—1972
Hawai'i Agricultural
  Company—1972
Pepe'ekeo Sugar
  Company—1973
Hāmākua Mill Company—1974
Kohala Sugar Company—1975
Laupāhoehoe Sugar
  Company—1979
Honoka'a Sugar
  Company—1979
(Second) Puna Sugar
  Company—1982
Ka'ū Sugar Company—1986
Davies Hāmākua Sugar
  Company—1984
Mauna Kea Sugar
  Company—1994
Hāmākua Sugar
  Company—1994
Ka'ū Agribusiness
  Company—1996 (sugar)

by as much as 25 percent. Despite great efforts to eradicate the disease at the Makiki site, including burning the cane and fumigating the soil, seven months later it was found on the O'ahu Kunia substation and on O'ahu Sugar Company. Though the infected cane was manually destroyed and fields were plowed, it was too late; wind-borne spores of the disease were already widely distributed. The disease was soon reported on Kaua'i (1973), Maui (1974), and Hawai'i (1979). Fortunately, it never became a serious problem on

Hawai'i because the moist environment was not conducive to smut development. On leeward plantations, it was controlled by hot water treatment of seed pieces, plowing out susceptible varieties, and planting new smut-resistant, high-yielding varieties developed by the Experiment Station.[13]

Despite financial challenges, the Experiment Station continued its support of the industry, and in 1975 its staff moved to a new building in Aiea. The four-story concrete structure housed all the laboratories, the offices, and the library that had taken up more than 20 buildings at the original site in the Makiki District of Honolulu. The library housed over 88,000 volumes in addition to the project files and crop files dating back to the origin of the Experiment Station. In 1981 Don Heinz, director of the Experiment Station, reviewed recent contributions of the Experiment Station to higher plantation yields. Among them were the production of improved varieties of cane; registration and use of Polado (glyphosate) as a cane ripener; registration of an insecticide (Amdro) for ant control in drip-irrigated fields; development of parallel-ridge drip irrigation tube for control of ant damage to drip tubes; and the success of agriculture and factory training programs for plantation personnel. Nevertheless, continued economic pressures on the industry required the Experiment Station to cap expenditures wherever possible and pursue income-producing outside contracts. In response, in 1982 the state legislature provided $3 million in financial assistance to the Experiment Station, allowing it to continue "to help Hawai'i's plantations find ways to cut costs while improving productivity."[14] Funds were also allocated for diversified crop research and for studying potential by-products of the sugar industry.

By 1985 the industry had begun to increase yields of sugar per acre and reduce production costs by planting more productive varieties, converting more acres from furrow to drip irrigation, and improving fertilizer, ripening, weed control, and other field and factory practices. Compared with the high point of 1982, Hawai'i's costs of production in cents per pound of sugar had come down by 4.3 cents, or 16 percent, in just three years. Though Hawai'i had been the highest-cost U.S. sugar producer in 1980, by 1984 it was the second lowest, higher only than Florida. In fact, Hawai'i had lower growing costs than any other state, though its processing costs were higher than Florida's.[15] Industry-wide record yields were achieved in 1986.

The industry came under public pressure in the mid-1980s to reduce its use of pesticides (almost entirely herbicides), causing the Experiment

Station to educate the public about the need for agricultural chemicals and increase chemical safety programs for industry personnel.[16] In 1988 the Experiment Station began micro-propagation of plants for transplanting. This technology was expected to speed the introduction of new and improved varieties because tissue culture could be used to multiply new varieties far more quickly than traditional seed fields. In addition, fields used to produce seed cane could be returned to commercial production. Three plantations requested the tissue-cultured planting materials, and by 1989 the Experiment Station was producing between 22,000 and 25,000 plants per week at a cost of 7.5 to 8.0 cents per plant.[17]

During the 1980s the state legislature encouraged the Experiment Station to conduct more research on agricultural diversification and by-products that could be derived from sugarcane. Often working in cooperation with the University of Hawai'i, the station carried out studies on cocoa, coffee, alfalfa, and tropical grasses for livestock feed, hybrid grass seed production, biomass for energy, and cover crops for soil erosion control. By-product research included ethanol from molasses, industrial chemicals from bagasse, pharmaceuticals derived from sucrose, and increased fiber production for energy or other products.[18]

Despite broadening its research portfolio, the Experiment Station remained under continuing economic pressure during the late 1980s and early 1990s. In response, it reduced staff, sought external funding, and sold its 40-acre quarantine station on Molokai, using the funds to supplement its budget. However, these measures were not enough, and in 1996, after 100 years of voluntary support by the sugar industry, the Experiment Station of the HSPA was renamed the Hawaii Agriculture Research Center (HARC). The name change reflected the continued movement to more research on diversified crops, but the organization remained a nonprofit 501(c)(5), not eligible for tax deductible donations but retaining the right to continue lobbying efforts for the sugar industry. Eight sugar companies were still operating in Hawai'i at that point, producing 437,261 tons of raw sugar, and the HARC board of directors remained in the hands of the sugar industry. Consequently, sugarcane breeding was still a primary activity, but other activities focused on coffee, forestry, and seed production, including production of transgenic Rainbow and Sunup papaya seed. A second phase of the transition occurred in 2009, when HARC changed to a 501(c)(3) education and research institute, making tax deductible donations possible but disallowing lobbying efforts. In 2010 HARC had ongoing research programs on sugarcane, coffee, papaya, taro, forestry, and anthurium.[19]

In their history of the HSPA Experiment Station, Heinz and Osgood[20] summarized its accomplishments as follows:

Over its 100 year history, the Experiment Station helped the sugarcane industry increase its average yield from 5.14 tons sugar per acre in 1908–1909 to 8.06 tons of sugar per acre in 1946 and to 12.47 tons of sugar per acre in 1986. Some of the highlights of the research that led to this high yield and to a better understanding of the biology and chemistry of the sugarcane plant were:

1. Development of high-yielding, pest- and disease-resistant varieties.
2. Identification and introduction of natural enemies to control insect pests.
3. Identification of herbicides to control weeds, fungicides to control seed rotting diseases, and plant growth regulators to increase sugar content and sugar yields.
4. Development of fertilizer practices for optimization of biomass and sugar yield in the biannual cropping cycle that differentiated Hawai'i from other cane growing regions.
5. Development and refinement of drip irrigation technology to optimize the distribution and conservation of water.
6. Mechanization of planting and harvesting operations eliminating the difficult hand operations associated with sugarcane agriculture. In the 1930s it took 50,000 workers to produce one million tons of raw sugar; in 1986, 5,400 workers produced the same amount.
7. Development of an industry environmental program including facilities for laboratory analysis of potential chemical residues.
8. Notable contributions to plant science such as elucidation of the C-4 pathway of photosynthesis, tissue culture, and meristem culture.
9. Participation in the development of the technology and methods of genomics to map the sugarcane genome.
10. Development of sucrose-based chemicals to make such products as thermally stable plastics, epoxy resins, and anticancer drugs.

It is a fair assumption that if not for the Experiment Station, many sugar companies would not have survived as long as they did.

Sugarcane Management

By the turn of the twenty-first century, the management of sugarcane in Hawai'i had adupted to the almost exclusive use of drip irrigation. Almost all aspects of growing the crop had been adapted in some way to this new and much more water- and labor-efficient method of irrigation. This section describes the industry's adaptation of drip irrigation to its needs during the 1970s–1990s and the impacts that it had on other sugarcane management practices.

Drip Irrigation

By far the most important crop management development in Hawai'i's sugar industry after 1970 was the conversion from furrow to drip irrigation. Early work on irrigation systems using perforated tubes to apply water in small amounts directly to plant root systems was conducted by Dr. Symcha Blass in Great Britain soon after World War II. The Experiment Station of the HSPA initiated work on this concept in 1959 by studying soil-wetting patterns from 2-inch polyethylene tubes with one-eighth-inch orifices. Similar work was conducted in 1965 using smaller diameter orifices. In 1970 representatives of HSPA visited Dr. Blass in Israel, where he had relocated, and Experiment Station engineers installed a subsurface irrigation system at Kunia. The experiment compared the growth of cane irrigated with half-inch polyethylene drip irrigation tubes of three different orifice sizes and buried at 12 and 18 inches. Preliminary results demonstrated good cane growth and the ability to precisely apply irrigation water, but challenges were also evident, including the need for good filtration systems and chemical treatments to prevent clogging of orifices by sediment, microbial films, and chemical precipitates.[21]

The Experiment Station trumpeted that it might have made "the long-awaited breakthrough in irrigation with the initial apparent success of trickle (or drip) and subsurface systems." As a result, Olokele Sugar Company began rapidly installing both drip (on the soil surface) and subsurface (buried) systems; by 1972 HSPA Experiment Station engineers could report that "subsurface or drip irrigation systems appear to overcome the inherent disadvantages of furrow irrigation and, to a degree, sprinkler systems because they can continuously 'feed' the cane roots through a network of small tubes installed along the cane lines."[22]

That same year three basic types of drip tubes were under evaluation. The simplest was a half-inch tube with orifices that were punched, drilled, or burned with a laser. The problem with this design was that flow rates along the tube had to be relatively high and pressure losses resulted in decreasing flow rates along the length of the tube. This flaw could be overcome by using larger diameter (three-quarter-inch) tubes operating at lower pressures. A third type of tube—and the one that became the favorite of the industry—had dual chambers: one that operated at higher pressures (~15 psi) and conveyed water along the tube, and a smaller low-pressure chamber (~1–2 psi) that distributed the water to the soil through tiny orifices spaced about 2 feet apart. The advantage of the dual-chamber tube was that it had uniform flow rates, even with very long tubes (up to 800 feet), and it could apply water at very low rates, which reduced the cost of pipelines needed to convey water to the fields. Experiment Station scientists were optimistic, predicting that "these systems will be installed on approximately 106,000 acres of furrow-irrigated land within the next 15 to 20 years."[23]

The industry had installed 300 acres of drip irrigation by 1972. Another 1,200 acres were installed in 1973, and in 1974 Experiment Station engineers claimed that "the Hawaiian sugar industry is rapidly converting furrow-irrigated lands to drip irrigation." Drip systems installed on or near the soil surface were destroyed at harvest, while subsurface systems were expected to last for several crops. Field trials had demonstrated that "gross water applications are considerably less than for furrow irrigation and somewhat less than for sprinkler irrigation." This savings was attributed to drip irrigation efficiencies of at least 80 percent compared with normal furrow irrigation efficiencies of 30–40 percent and sprinkler irrigation efficiencies of 65–75 percent.[24]

Systems were sized to apply approximately 80 percent of pan evaporation (during the summer peak-use period) in a 12-hour period (about 0.32 inches per day). Experiments were under way to determine if irrigation could be applied continuously through drip and subsurface systems during the peak-use period to reduce the cost of main pipelines conveying water to the fields. Because the cost of tubing was significant, especially with drip systems that were destroyed at harvest, most plantations were using "pineapple planting" in which the drip tube was installed between two lines of cane (3 feet apart with 6 feet between the paired rows). Since the drip line in the pineapple planting configuration was between rather than directly above the line of seed pieces, in some soils water did not move far enough laterally to assure that the soil surrounding the seed pieces was wet enough to stimulate germination and early growth of the young root and shoot systems. In such cases plantation personnel often placed the drip line directly over the line of seed pieces, applied irrigation to wet that line, then moved the line over the other line of seed pieces to irrigate it, finally placing the line midway between the lines after germination was assured.[25]

Plugging of tube orifices was soon identified as the greatest single problem associated with subsurface and drip irrigation, even after screen and sand filters had reduced the solids in the irrigation source water to less than 60 micrometers and 2–3 parts per million (ppm). Plugging was caused by a combination of factors, including growth of algae at the orifice, mats of aggregated solids, and salt crystals at the orifice. Periodic flushing of the tubes (by opening their ends during irrigation) and chemical treatments to prevent algal growth promised to minimize these problems. Another problem involved ants that chewed the emitters, increasing the size of the holes and producing uneven distribution of irrigation water along the line. This problem was solved by ingenious Experiment Station entomologists who discovered that the ants had to pivot around and around the emitter in order for their mouth parts to enlarge the hole. They then worked with tubing manufacturers to produce tubes with tiny plastic ribs on either side of the orifices to prevent the ants from positioning their heads to enlarge the orifices.[26]

Major investments were required between the mid-1970s and late 1980s to make the conversion, including installation of filtration systems to provide clean, sediment-free water that would not plug the orifices of drip tubing. Many of the plantation ditches that had carried irrigation water for a century were removed, and pipelines were constructed to distribute water from the filter systems to the fields. Finally, drip irrigation tubing had to be installed and connected to the water distribution system in the fields after each harvest. Despite the costs of the conversion, adoption of drip irrigation produced numerous benefits, including (1) greatly reduced labor needed to apply furrow irrigation and fertilizers; (2) elimination of tillage operations needed to create and maintain irrigation furrows; (3) substantial reduction of inter-row cultivation for weed control; (4) conversion of land in irrigation ditches to cropland; (5) greater efficiency of irrigation application; and (6) increased yields by providing irrigation water and soluble fertilizers precisely and directly to the cane roots. Wailuku Sugar Company reported "a material increase in cane and sugar yields and a 30% improvement in water efficiency." In addition, "very little manpower is required to irrigate after the system is installed." However, at Pioneer Mill Company and Wailuku, rodents chewed holes in drip tubing near the edges of sugarcane fields adjacent to rock piles and waste areas, especially during periods of hot, dry weather.[27]

As of 1980 drip irrigation was in use on about 60,000 acres, just over half of the state's irrigated sugarcane, and plantations were reporting that drip-irrigated cane seemed to have slightly higher yields than even well-managed furrow-irrigated cane. But because of the short drip irrigation intervals, researchers feared that cane roots would be concentrated near the soil surface, and the crop would be very susceptible to any lengthening of irrigation intervals due to temporary water shortages or equipment failures. To evaluate this issue, an experiment was conducted on Lihu'e Sugar Plantation to measure the sensitivity of drip-irrigated cane to short periods of water stress. Results clearly showed that cane growth began to decrease after only 50 mm (2 inches) of soil water had been used by the crop since the last irrigation. In addition, stalk elongation began to decrease well before other indicators of drought stress were evident, such as leaf rolling or yellowing, changes in false-color infrared reflectance, changes in leaf water potential, or a reduction in the rate of water extraction from the soil. These results drove home the importance of good drip irrigation interval control and system maintenance to maximize cane growth.[28]

In addition, several indicators of cane growth (among them number and weight of internodes, leaf area per stalk), as well as tons of cane and sugar per acre, increase linearly (or nearly so) for drip-irrigated cane irrigated at pan factors between 0.35 and 1.2. ("Pan factor" is a number that reflects how much water a plant uses in a given time period, which helps growers determine when irrigation of a crop is necessary; it is calculated as rain + irrigation / pan evaporation.) These results were very similar to the results of earlier furrow- and sprinkler-irrigated experiments—namely, that maximum cane and sugar yields require the cane crop receive enough water to meet the evaporative demands of the atmosphere. Other studies indicated that the ratio of thermal infrared to red reflectance of the cane canopy could be used to detect areas of fields receiving less-than-optimum irrigation.[29]

Several studies during the 1980s demonstrated that in order to obtain the maximum benefit from drip irrigation, it was necessary to eliminate other factors that could limit crop growth. For example, cane yields were not as responsive to irrigation if nitrogen and potassium fertilizer rates were inadequate. Yields were more responsive to phosphorus fertilization when drip irrigation was adequate than when irrigation was inadequate. Drip-irrigated sugarcane also responded positively to breaking up compacted surface soils by disking, primarily because soil compaction reduced the quality of the juice. Finally, cane on poorly drained soil with a high water table grew poorly and did not respond to variation in drip irrigation amounts.[30]

When water supplies were inadequate to fully irrigate all the cane on a plantation, managers were faced with the issue of whether to apply more irrigation to younger or older cane. A review of irrigation experiments demonstrated that for furrow- or sprinkler-irrigated cane, the time at which drought stress occurred had little effect on the relation between water use and cane yield. This conclusion was later confirmed with drip-irrigated cane in an experiment that found little difference in final yields of cane and sugar whether drought was imposed in the first or second year of growth.[31]

By 1986 almost 91,000 acres of sugarcane were under drip irrigation, and only 20,000 acres remained in furrow or sprinkler irrigation. As of 2000, about 92 percent of all sugarcane acreage in Hawai'i was drip irrigated, providing precise control of water and fertilizers. Weather data were used to estimate potential evapotranspiration (averaging 0.2–0.3 inches per day), and drip irrigation amounts were carefully controlled to maintain soil water near field capacity.[32]

Planting

Before planting, fields were normally subsoiled (deep plowed) to a maximum depth of 24 inches and/or plowed with 36- to 42-inch-diameter discs (see Figure 5.1). This broke up compacted layers and left a flat and finely textured seedbed ideal for drip irrigation. Sugarcane seed pieces about 14 inches long were mechanically planted in furrows 6–10 inches deep and covered with 2 inches of soil. By 2000 ratooning was uncommon, and over 70 percent of the fields were tilled and planted every two years. This allowed the plantations to reduce soil compaction caused by harvesting, improve soil aeration and drainage, change the cultivar when needed, incorporate soil amendments and residues, minimize insects and other pests, and more easily eliminate perennial weeds.[33]

Weed Control

Weed control was expensive, costing $80 to $150 per acre per crop of 24 months, but, when uncontrolled, weeds reduced cane yields and sugar recovery in the mill. Highly skilled weed control specialists managed herbi-

Figure 5.1. Soil preparation before planting. Photo courtesy of HARC

cide applications. Problem weeds included guineagrass (*Panicum maximum*), alexandergrass (*Brachiaria plantaginea*), swollen fingergrass (*Chloris radiata*), Aiea morningglory (*Ipomea triloba*), and peria (*Momordica charantia*).

Soil-applied herbicide rates were higher in Hawai'i than in the U.S. mainland because tropical soils, with their high iron oxide contents and large surface areas, tend to adsorb and deactivate herbicides. Also, it takes several months for the newly planted sugarcane's leaf canopy to shade the soil, so it needs extended protection from weed competition. Herbicides accounted for 97 percent of all sugarcane pesticide use, and five herbicides (atrazine, diuron, pendimethalin, glyphosate, and ametryn) represented 80 percent of the total pesticide usage.[34]

Prior to preplant tillage or ratooning, perennial weeds like the aforementioned guineagrass (*Panicum maximum*), as well as napiergrass (*Pennisetum purpureum*), were sprayed with glyphosate or hexazinone. After the drip system had been installed and planting or replanting was complete, a preemergence herbicide mix—usually pendimethalin and atrazine—was broadcast by tractor, or often, when soils were too wet for tractors, by air. Aerial application of herbicides was discontinued in the 1990s at HC&S but was continued on some other plantations. Post-emergence applications were used to kill weeds not controlled by the pre-emergence application. Field edges sometimes received additional applications (usually glyphosate and/or 2,4-D) to keep the drip irrigation systems free of weeds for easy maintenance and service.[35]

Fertilizers and Soil Amendments

Fertilizer amounts were usually determined based on soil and crop tissue analyses, planting and harvest dates, crop growth stage, and weather. Typically, the soil was sampled prior to planting and analyzed for pH, phosphorus, potassium, calcium, and magnesium. Other elements were analyzed in some fields in which deficiencies or toxicities were suspected. Lime was incorporated into acid soils (pH less than 6.0) prior to planting, but when calcium was deficient on soils with neutral or higher pH, gypsum was applied, especially on Maui. Irrigation was usually applied within 24 hours of planting or replanting, and initial applications of nitrogen, phosphorus, and potassium were applied via the irrigation water within two weeks of the first irrigation. All other fertilizers were applied with water through the drip system, with phosphorus, sulfur, and calcium applied in one or two applications early in the crop. Leaf sheaths and blades were monitored for nutrient deficiencies during the grand growth period, usually from six to nine months of age. The last nitrogen was applied about ten to twelve months after planting for a 22- to 24-month crop. Later application could stimulate production of secondary shoots and rapid stem elongation, both of which reduced the storage of sugar and as a result lowered sugarcane quality.[36]

Flowering Control

The apical meristem of a cane stalk stops producing leaves and initiates a panicle of flowers in response to increasing night lengths during September. The panicle grows, and the seed heads, also called "tassels" or "arrows," emerge from the whorl of leaves at the top of the stem about 45 days later. The percentage of stalks that produce panicles varies with variety, hours of daylight, temperature, and age of the crop during the critical autumn months, and flowering in the first year of a two-year crop reduces cane growth and the sugar content at harvest due to cessation of stalk growth and initiation of many lateral shoots that are still immature at harvest. The only nonchemical methods to control flowering were to cut back the cane and to withhold irrigation to stress the crop prior to flower initiation. The desiccant Diquat sprayed over the cane foliage was somewhat effective for reducing flowering and was registered for this use; however, the growth regulator ethephon was shown to both inhibit flowering and increase cane tonnage. As a result, in 1998 ethephon was applied to about 7 percent of sugarcane acreage, usually 6-month-old to 15-month-old cane, to inhibit flowering. Some

Hawaiian varieties are notable for their low flowering—for example, H65–7052, a prominent variety on Maui, did not normally flower in commercial fields.[37]

Ripening

Unlike some subtropical sugarcane producing areas, Hawai'i does not have a sufficiently cool, dry period to induce good natural ripening. As a result, the industry used a three-pronged approach to increasing sugar storage in mature stems, including depleting the soil and crop of nitrogen, withholding irrigation, and applying sublethal rates of the growth regulator glyphosate (Polado) as a chemical ripener. All three techniques reduce stem elongation more than they inhibit photosynthesis, thereby enhancing sucrose accumulation in the cane stem. As noted earlier, nitrogen fertilization was terminated from nine months to a year before harvest to allow the crop to enter a period of nitrogen stress. Four to five months prior to harvest, irrigation was gradually reduced to slow cane elongation and "harden" the crop for final ripening. Early work by the Experiment Station on chemical ripeners resulted in glyphosine being registered by EPA as the first ripener for use in Hawaiian sugarcane. This work was followed by the discovery that the herbicide glyphosate (a close relative of glyphosine) was effective at lower rates than glyphosine and provided a greater assurance of a ripening response. Used at sublethal rates as a chemical ripener, glyphosate (Polado), became widely used in Hawai'i. Applied aerially at 0.25–0.5 lb active ingredient per acre six to eight weeks prior to harvest, it inhibited apical growth, enhanced sugar accumulation, and improved juice purity (the ratio of sucrose to soluble impurities in the extracted juice). After Polado application, irrigation was eventually terminated to dry the crop and facilitate pre-harvest burning. The leaf sheath moisture, sugar content, and/or blade nitrogen concentration were monitored weekly until harvest to determine the optimum time to harvest.[38]

Harvesting

Weather permitting, fields were burned and within 24 hours the cane was pushed into piles with push rakes (see Figure 4.12) or into windrows with V-cutters (see Figure 4.14). The cane was then loaded onto trucks using hydraulic cranes (see Figure 5.2). The industry had been experimenting with combine harvesters (see Figure 5.3) for many years to facilitate harvesting

Figure 5.2. Hydraulic grab harvester. Photo courtesy of Lance Santo

Figure 5.3. Experimental cane harvester. Photo courtesy of HARC

of green, unburned cane, but commercial use of this equipment remained uneconomical because of the complex tangle of lodged cane stalks, heavy tonnage of the two-year-old cane, and in some cases the presence of rocks. Tractor-trailers delivered the cane to the factory on roads maintained by each plantation. In the trailer the cane rested on metal chains that were attached to a metal beam called a "strongback" on one side of the trailer. When the tractor-trailer arrived at the factory, a crane lifted the beam, which raised the cane and rolled it out of the trailer over a retaining wall (see Figure 5.4).

Figure 5.4. Crane unloading cane at mill by raising strongback attached to chains. Photo courtesy of Lance Santo

The piles of cane were then washed with water to remove soil and other extraneous material before being milled. Solids in the wash water were allowed to settle, and then the water was used for irrigation; the sediment, rich in organic matter, was used to amend fields with poor soils.[39]

Insects, Diseases, and Rats

Pests, including insects, diseases, and three species of rats, continued to be a chronic problem for sugarcane growers. The Polynesian, Norway, and black rats fed on stalks, attracting sugar cane borers and fungal diseases to the wounds they caused. The rats had survived the introduction of the mongoose and trapping, and they nested in rock piles, gulches, and waste lands near cane fields. Control of perennial grass weeds in fields and on roadsides was necessary to limit rat nesting sites. In addition, anticoagulants and zinc phosphide were used as rodenticides, although the anticoagulants were discontinued due to secondary poisoning concerns. Feral pigs also chewed stalks, uprooted plants, and damaged drip irrigation systems. Control methods included electric fencing, hunting, and trapping.[40]

In Hawai'i classical methods for biocontrol of sugarcane insect pests had been quite successful since the 1890s, controlling more than 10 damaging species. As a result no insecticides were applied in the state's commercial sugarcane fields. Experiment Station entomologists imported natural enemies, usually parasitoids, of the pests from other sugarcane growing regions.

After quarantine and establishing that no other crops or native plants would be attacked, the biocontrol agents were mass reared and released at sites infested with the pests. Repeated releases were sometimes needed to establish viable parasitoid populations, but adequate levels of control often resulted without reintroduction.[41]

During the 1980s and 1990s the three most important insect pests were the New Guinea sugarcane weevil, the lesser cornstalk borer, and the yellow sugarcane aphid. The weevil had been a chronic and persistent pest since the 1880s, and the tachinid fly, a parasite of the weevil, was introduced from New Guinea in the early 1910s and again in early 1970s. By 2000 the fly was providing substantial control of the weevil. The borer was the most destructive pest of young sugarcane, killing shoots up to six weeks of age and damaging thousands of acres on Maui between 1986 and 2000. It was more abundant in dry, hot areas, but two parasitoids were introduced and, in combination with more resistant cultivars and frequent irrigation prior to close-in, they had reduced borer damage. The aphid became established in Hawai'i sugarcane fields in 1989. Though parasitoids from Europe had been introduced in 1991–1992, they had not produced control, and in 2000 the search for effective biological control agents continued.[42]

The major sugarcane diseases faced by the industry were pineapple disease, smut, rust, eyespot, ratoon stunting disease, leafscald, and yellowleaf syndrome, a virus. Pineapple disease is caused by a fungal pathogen that enters the cut ends of seed pieces, killing the buds and preventing germination. It was controlled by dipping seedpieces in a heated propiconizole (Tilt) solution prior to planting. Smut is a fungal disease with spores that can be spread though infected seed pieces. Fungicides were ineffective, but the industry kept losses to a minimum, primarily by developing and planting resistant cultivars, but also using uninfected planting material and dipping seed pieces in hot water prior to planting. Rust is a fungal disease that produces small, rust-colored leaf lesions that reduce photosynthesis and stunt crop growth, but the industry achieved acceptable control by breeding cultivars with good rust resistance. Eyespot is a fungal disease that produces eye-shaped leaf lesions. Spread by airborne spores, it produces a toxin that causes severe plant necrosis, stunting, and death in susceptible cultivars. But by 2000 the disease was no longer prevalent because resistant cultivars had been planted for many years. Ratoon stunting is a bacterial disease that produces no visible symptoms other than stunting. Though no effective pesticides or resistant cultivars were available, new seed fields were established with seed pieces that had received a two-hour hot water treatment to eliminate the bacterium. If the disease was suspected in a seed field, harvesting knives were sanitized with sodium hypochlorite bleach before moving to a new field. Leafscald is a bacterial disease that produces colorless streaks on leaves and can cause severe yield loss. It is spread by infected seed pieces, but it was controlled with resistant cultivars. Yellow-leaf syndrome is caused by a virus that produces severe leaf yellowing and is transmitted by the sugarcane aphid; symptoms varied among cultivars and cultural practices, but breeding resistant cultivars provided adequate control.[43]

Diversified Agriculture

Sixteen Hawaiian sugar mills closed in the 1970s and 1980s, so by the late 1980s the HSPA was searching for alternative business models that could reduce industry dependence on federal price support programs and a volatile world sugar market. To stimulate thinking about industry diversification, the 1989 Annual Conference of Hawaiian Sugar Technologists focused on developing new products from cane, experimenting with alternative crops, and broadening markets. John Couch, chairman of HSPA, pointed out that the gasoline crisis of the early 1970s had spurred research for production of alternative fuels like ethanol from sugarcane. Couch urged the industry to follow Brazil's lead and produce ethanol from molasses. Bagasse could be used to make paper, and hemicellulose from bagasse could be used to produce furfural, an industrial chemical with multiple uses. Lignin, also from bagasse, could be used to produce phenolic resins used in plywood and other building materials. Couch predicted that "twenty years in the future, Hawaiian cane farmers still will be providing sugar for human consumption. But they also may be producing alternative energy, paper pulp and even biodegradable plastic from sugarcane." He pointed out that while the industry had been selling electricity and molasses for years, it was now "experimenting with alternate crops ranging from rubber to herbs for fragrances."[44]

However, despite continued research at the Experiment Station, no major diversification of the sugar industry occurred during the 1990s, and by the end of the decade it was clear that most of Hawai'i's sugar companies would soon be out of business. Only two sugar companies were still operating at the close of 2000: Gay & Robinson on Kaua'i and HC&S on Maui. The industry had declined from over 240,000 acres at its peak to less than 50,000 acres, and its collapse freed up large amounts of agricultural land, much of it with well-developed, although rapidly deteriorating, irrigation systems. The Hawai'i Department of Agriculture saw an opportunity to expand diversified agriculture to replace imported fruits and vegetables, as well as to compete in niche markets on the U.S. mainland and overseas. This optimistic assessment was based on the growing value of diversified agriculture, from about $90 million in 1980 to almost $280 million in 2000—an amount that partially offset the decline of sugar and pineapple sales from about $460 million to about $160 million over the same period. However, diversified agriculture needed irrigation water, and most of the plantation irrigation systems had been abandoned and were falling into disrepair. The Hawai'i State Legislature commissioned a plan to inventory the state's irrigation systems; establish the needs, criteria, and priorities for their rehabilitation; and develop plans for their rehabilitation and long-term maintenance. The resulting plan identified $100 million in total costs needed to rehabilitate 10 irrigation systems, not counting the East and West Maui systems that were still in use by HC&S.[45]

In fact, the Department of Agriculture's optimism was not misplaced. Total crop production receipts for Hawai'i increased from $431 million in 2000 to $630 million in 2011. Several categories of crops increased substantially from 2000 to 2011, most notably seed production (from $38 million to $247 million), but also fruits and nuts (from $84 million to $100 million) and cane for sugar (from $62 million to $78 million). By far the biggest factor in the expansion of diversified agriculture in Hawai'i was the rapid expansion of production of parent seed by five major seed companies.[46]

Biomass for Energy

Located in an island state with very high energy costs, the Hawai'i sugar industry has, throughout its history, used bagasse to produce the energy it has needed to power its sugar factories and pump irrigation water. Excess power was first used in the industry to electrify plantation communities. HC&S's Pu'unēnē factory generated the electricity for the islands' first electric lights, and the company has long generated a significant fraction of the electricity for Maui Electric Company, both with its bagasse-fired boilers and its hydropower plants. During World War I Maui Agricultural Company had fermented the sugars in molasses and distilled ethanol to power its trucks, automobiles, and tractors. In the 1960s HC&S partnered with Seagram's to produce rum at Pu'unēnē. It is no wonder, therefore, that when energy prices rose or sugar prices declined, HC&S and other sugar companies considered diversifying to sell even more energy, as either electricity or liquid fuel. Spurred by the Brazilian sugar industry's success in producing transportation ethanol as well as sugar, the increasing production of corn-based ethanol in the United States, and government mandates to blend ethanol in gasoline, the Hawai'i sugar industry considered biofuels as a possible key to its survival.

Several studies in Hawai'i evaluated the potential for using biomass from trees and grasses (including sugarcane) to produce energy on a commercial scale. A seven-year study compared biomass production by tree and grass crops at five diverse locations across the state.[47] Among the trees tested, *Eucalyptus* species demonstrated good yield potential in high-rainfall areas, and *Leucaena* was well adapted to low-rainfall areas, but supplemental irrigation could greatly increase its yields. In all but one location (Honoka'a), sugarcane or *Pennisetum purpureum* (banagrass) produced greater dry biomass yields than the tree species, producing an average of 13.8 tons of dry matter per acre per year over the five sites evaluated, compared with an average of 9.2 tons for trees. Maximum yields for the banagrass approached 20 dry tons per acre per year at Hoolehua, Moloka'i. The maximum yield for trees was 14.2 tons per acre at Honoka'a. Despite these good biomass yields, the study concluded that in the early 1990s it was not economically feasible to produce, deliver, and process stand-alone biomass fuels from either trees or grasses to generate electricity, and "the best biomass option for Hawai'i [was] to continue to produce a higher-value product such as sugar and produce energy as a by-product."

A study[48] of the potential for biomass-ethanol production in Hawai'i found that a large fraction of the agricultural land that had gone out of sugarcane production was suitable for production of biomass ethanol crops. Sugarcane production had decreased from about 255,000 acres in the 1930s to about 70,000 acres in 1999. Most of these abandoned lands remained fallow, and much of the irrigation and transportation infrastructure remained in place, providing "an unprecedented opportunity for establishing a new agri-energy industry." The study evaluated the potential for growing sugarcane, banagrass, and *Eucalyptus* and *Leucaena* trees on three sites: Waialua Sugar Company lands on O'ahu (12,000 acres), HC&S lands in Wailuku and Makawao on Maui (20,000 acres), and Hāmākua Sugar Company lands on the Hilo Coast of Hawai'i (more than 20,000 acres). All three sites had adequate surface and groundwater resources, transportation networks, and idle industrial sites available to house electrical generation or ethanol production. In addition, all three islands had significant amounts of municipal solid waste and agricultural waste from the sugar, pineapple, macadamia, and forestry industries to complement dedicated biomass energy production. The study concluded that the three sites had great potential for producing biomass energy crops at costs ranging from $54–85 per dry ton. If a suitable commercial-scale lignocellulose conversion technology could be developed, ethanol could be produced for $0.55–1.07 per gallon. However, no proven, large-scale conversion technologies could be identified, and the U.S. recession of 2001 discouraged major projects involving conversion of idle sugarcane lands to bio-energy crops.

However, strong interest in increasing bioenergy production by the Hawai'i sugar industry continued in some circles beyond the 2001 downturn. The Hawai'i Agriculture Bioenergy Workshop in 2006 included a number of presentations about how agriculture might produce bioenergy for electricity and liquid fuel. The only two remaining sugar companies, Gay & Robinson on Kaua'i and HC&S, were both developing plans to increase their production of bioenergy. Gay & Robinson considered a system in which sugarcane juice would be fermented and distilled to ethanol, while bagasse would be converted to syngas to be reformed catalytically to produce ethanol, as well as to drive a gas turbine to generate electricity. If this technology, at the time in the pilot plant phase, proved scalable and could be installed at the Gay & Robinson Plantation, it was projected to produce net revenue of $7,388 per acre, compared to $2,708 for traditional sugar production and $3,375 for ethanol produced with traditional technology. But the estimated cost of implementing such a project was $90–100 million, and Gay & Robinson ceased operations in 2010 before the conversion technology could be proved and a biofuels project could be implemented.[49]

HC&S also considered several alternative approaches to produce biomass energy. In 2005 it had generated approximately 212,000 megawatt hours (MWh) of electricity, 87 percent from bagasse and 13 percent from hydropower. About 55 percent was used by the company and 45 percent was sold to Maui Electric Company, providing roughly 7 percent of MEC's need for firm power. If HC&S could eliminate pre-harvest burning in order to harvest and burn its cane trash, it could almost double its electricity production, but tests showed that the costs of harvesting, hauling, and grinding the cane trash were greater than the value of the additional electricity generated. A second possibility was to ferment its molasses to generate approximately 5 million gallons of ethanol per year. In addition, by fermenting a fraction of its raw cane juice to supplement the molasses, the company could produce more than 10 million additional gallons of ethanol. Major impediments to producing such large amounts of ethanol were the economics of production and safe disposal of the *vinasse,* effluent waste resulting from fermentation. Though the vinasse could be used as a partial replacement for fertilizer potassium, its odor, acid pH, high temperature, and high concentrations of salts and organic matter would require costly treatment. Because of such economic and environmental concerns, HC&S had not implemented such bioenergy projects by 2014.[50]

HC&S Responds to Economic and Environmental Challenges

In the late 1970s and 1980s, dramatic fluctuations in sugar prices and the uncertainty about long-term domestic price support programs put severe economic pressure on HC&S. In addition, from 1970 to 2000 the population of Maui grew from less than 39,000 to almost 118,000. Increasing urbanization and environmental awareness of the population resulted in growing public concern about the role of agriculture. These concerns led to protests about smoke from pre-harvest cane burning and from sugar mills. Concerns for the health of the East Maui watershed and the desire to expand taro production led environmentalists and taro producers to join forces and advocate leaving more water in the streams feeding the East and West Maui water diversion systems. They argued that HC&S was using its irrigation water inefficiently and wastefully, and more should be available for taro production and to maintain stream flows. Beginning in the 1970s HC&S responded to these economic and environmental issues by increasing the efficiency of its irrigation system, continuing to reduce the size and increase the efficiency of its workforce, installing equipment to eliminate factory smoke, and reducing the impact of pre-harvest burning on its neighbors.

Drip Irrigation

The early 1970s marked a turning point for HC&S and most of Hawai'i's irrigated plantations. HC&S management projected that, even if the domestic sugar program remained in effect, by the late 1980s the company would be unprofitable, largely the result of increasing costs of labor, machinery, and other inputs. Despite extensive research and implementation of more automated furrow irrigation systems, the labor associated with irrigation and land preparation to form and maintain irrigation furrows was high and getting higher. In response, and with the experience gained from large-scale sprinkler irrigation experiments, management decided to begin gradual conversion to sprinkler irrigation. Eliminating irrigation furrows and adopting "flat culture" was expected to increase sugar yields, improve the efficiency of agronomic practices such as weed control, and reduce labor and tillage costs. It was thought that these savings would outstrip the increased capital equipment costs needed to install sprinklers and the costs of energy to pressurize the sprinklers. A seven-year program was initiated in 1972 to convert 14,800 acres of HC&S cane fields to sprinkler irrigation. But only 3,919 acres of sprinklers had been installed when, in 1974, the company decided to convert instead to drip irrigation. Over the next dozen years HC&S converted almost all its acreage to drip irrigation (an expenditure of $30 million in 13 years), and it was the largest drip-irrigated farm in the United States. By 1990 the management of Alexander & Baldwin (A&B) could report that "if it were not for drip, HC&S might have folded some years ago."[51]

Drip irrigation had many advantages over both furrow and sprinkler systems. For example, cane irrigated with drip systems made use of most of the water applied, with 80 percent or more being transpired by the crop for growth compared with only about 50 to 60 percent for furrow irrigation. In addition, water could be applied frequently in small amounts, virtually eliminating excess soil water and losses to deep percolation immediately after a furrow irrigation round. Frequent applications also eliminated the water stress that often occurred prior to the next furrow irrigation round. Small amounts of water applied frequently allowed drip irrigation on almost all soils and slopes, even those with low infiltration rates. Drip systems also minimized the problem of irrigating with brackish well water. With furrow or sprinkler systems the soil dried between irrigations, producing salinity-induced water stress between rounds. But with drip systems the soil remained moist and irrigation amounts could easily be increased when needed to force salts below the root zone. Other advantages included the ease of accurately applying soluble fertilizers with irrigation water though the drip system. Also, since drip irrigation systems worked at much lower pressures (around 15 pounds per square inch [psi]) than sprinkler systems (about 70 psi), drip systems required much less energy than sprinklers. HC&S found that converting 3,919 acres from sprinklers to drip reduced pumping power from 3,145 horsepower operating 24 hours a day to 1,060 horsepower operating only 60 percent of the time. In addition, many of the cane fields below the canals feeding the drip systems could be pressurized by gravity. Finally, the

Figure 5.5. Sand filters for drip irrigation. Photo courtesy of Lance Santo

increased irrigation water-use efficiency of drip irrigation allowed water to be transferred to chronically water-short areas of the plantation.[52]

As might be expected, large-scale conversion from furrow to drip systems was not without challenges. Water from the main plantation canals was brought to the plantation's reservoirs. From there it was pumped to arrays of sand filters, which used fine sand to remove silt and clay particles from the water (see Figure 5.5). These sand filter systems were automated to "back-flush" whenever the sand had trapped enough silt and clay particles to decrease flows through the filter, discharging the fine particles while retaining the clean sand within the filter.

From the sand filters the irrigation water was directed through main pipelines to the fields requiring irrigation. It was then conveyed through sub-main plastic pipes buried 3–4 feet underground to the edge of the field, where it rose to the surface in plastic "risers." These risers were connected to flexible drip irrigation tubing on the surface. Tiny holes (0.003 inches) located about every 2 feet along the tube emitted a continuous trickle of irrigation water. Some plantations installed one drip tube beside each cane row; but to save money, HC&S installed one line between two cane rows spaced about 3 feet apart. These paired rows were separated from adjacent pairs of rows by inter-rows about 6 feet wide in what was termed the "pineapple" spacing because it resembled the row configuration used in Hawai'i's pineapple industry.[53]

As more and more drip irrigation was installed at HC&S, systems and their maintenance improved. Stainless steel screen filters were introduced and improved, back-flushing of sand filters became automated, and drip system layouts were improved to facilitate planting and weed control. By the late 1980s drip irrigation system maintenance had become more efficient, with separate crews responsible for routine maintenance, repair of in-field systems after harvest, and repairs and upgrades of the HC&S system's more than 200 irrigation filter and control stations.[54]

Economic and Production Challenges

In response to the greater cane and sugar yields produced by conversion to drip irrigation, in 1979 HC&S began to expand the milling and sugar processing capacity at its Pu'unēnē mill. It added a fifth set of rollers to one of its two-mill trains, expanding its milling capacity by 35 percent, from 6,500 in 1979 to 8,800 tons of cane per day in 1984. It also added a new vacuum evaporation pan in 1980 and replaced the mill's original four reciprocating steam engines with eight steam turbines. The additional cane tonnage produced by drip irrigation generated an excess of bagasse, which was stored then burned in the off-season to generate electricity, replacing the oil normally used when the mill was not operating. The excess bagasse disappeared two years later, when a new 15 MW generator was installed, boosting Pu'unēnē's generation capacity to 35 MW and increasing HC&S's sale of electricity to Maui Electric Company.[55]

Despite these improvements in factory capacity, low sugar prices and increasing costs of production during the early 1980s almost brought HC&S to its knees. A drought from November 1980 to September 1981 reduced 1981 production from a projected 205,000 tons to 188,526 tons. Wet weather delayed the start of the 1982 harvest by two months and caused 37 rain-out days. To make matters worse, because of low sugar prices the C&H Refinery expected to be able to pay HC&S only $328 per ton of sugar in 1982. As its problems mounted, HC&S was facing a pre-tax loss of $12.8 million. The company responded to this disastrous turn of events by implementing a severe cost reduction program designed to save $7.2 million and reduce pre-tax losses to less than $4 million. The major elements of the program included over $2 million in savings from a four-week shutdown of operations, more than $2 million in fuel oil savings, and just under $1 million in savings resulting from rescheduling wage increases. As a result of bad weather and the four-week shutdown, 1982 production was only 165,900 tons of sugar, the lowest company production since the strike of 1959, and over 3,900 acres had to be carried over for harvest in 1983.[56]

Because of an increase in harvested acreage (due to the carryover from 1982) and better yields due to drip irrigation, the company produced a record crop of 214,806 tons in 1983. Despite the worst summer drought in its history, HC&S utilized drip irrigation to produce another record, 223,414 tons, in 1984; a near-record 219,468 tons in 1985; another record of 229,228 in 1986; and yet another of 232,718 in 1987. The 1987 record resulted from the plantation completing its conversion to drip irrigation, with the resulting water savings allowing it to expand its cane acreage by 4,000 acres. Sugar yields reached an all-time high of 14.7 tons per acre, and molasses production decreased because of better juice quality and higher sugar recovery.[57]

Sugar yields were very good in the mid-to-late 1980s, remaining above 13 tons per acre from 1986 through 1990 and exceeding 14 tons per acre in 1987 and 1989. However, they began a gradual decline in the early 1990s, falling below 13 tons per acre in 1991, 1992, and 1994, then dipping below 12 tons per acre in 1995, 1996, and 1997 (see Table 5.1). The declining yields caused management to seek answers, and in 1997 the HC&S Agricultural Research Department produced a comprehensive analysis that identified several possible problems. These included reduced East Maui Irrigation Company ditch flows, which (with the exception of 1991) fell below an average of 65 million gallons a day (mgd) from 1989 through 1994.[58]

The report also noted that since the conversion of the plantation to drip irrigation, the normal leaching of salts from brackish well water by excess furrow irrigation had doubtless decreased, a problem aggravated by low rainfall during the period 1991–1995. Insufficient leaching allowed soil salinity to increase, resulting in degradation of soil structure, poor infiltration, and lower yields. Of particular interest was the recognition that high concentrations of magnesium in Maui lavas produce high concentrations of magnesium in well waters. The well water, high in both sodium and magnesium, was causing several problems, including poor plant uptake of potassium and loss of soil structure. HC&S agronomist Mae Nakahata pointed out that both the problem and its solution had already been addressed on

Table 5.1. HC&S yields from 1970 to 2000

Year	Area harvested Acres	Sugar produced Short tons	Sugar yield Tons/acre
1970	14,938	192,530	12.89
1971	15,408	205,002	13.31
1972	15,647	193,933	12.39
1973	15,436	187,064	12.12
1974	15,013	181,970	12.12
1975	15,736	177,963	11.31
1976	15,085	185,405	12.29
1977	15,818	187,678	11.86
1978	15,665	170,166	10.86
1979	14,214	176,234	12.40
1980	15,356	188,004	12.24
1981	16,286	188,526	11.58
1982	13,653	165,900	12.16
1983	16,640	214,806	12.89
1984	17,047	223,414	13.11
1985	16,903	219,468	12.98
1986	16,525	229,228	13.88
1987	15,806	232,718	14.72
1988	16,490	229,388	13.91
1989	16,430	232,079	14.13
1990	17,089	225,555	13.20
1991	17,340	214,122	12.35
1992	15,715	193,388	12.31
1993	16,726	224,677	13.43
1994	16,457	206,217	12.53
1995	17,661	198,009	11.21
1996	17,185	201,019	11.70
1997	17,005	198,037	11.65
1998	17,024	216,188	12.70
1999	17,278	227,832	13.19
2000	17,266	210,269	12.18

Source: Hawaiian Sugar Planters' Association records.

Central Maui in the 1920s. A letter from Experiment Station director H. P. Agee to the manager of Maui Agricultural Company in 1927 had noted research in Central Maui showing that when levels of magnesium exceeded calcium in the subsoil, cane growth was very poor. Nakahata also observed that high concentrations of sodium and magnesium were associated with dispersion of soil aggregates, poor infiltration of irrigation water, and increased incidence of root rot. In addition, a number of HC&S fields were known to be deficient in sulfur, others had very acid subsoils, and in others the cane had symptoms of calcium deficiency. All these problems could be overcome by applying gypsum (calcium sulfate), then using excess drip irrigation to leach the calcium into the subsoil and increase its ratio of calcium to magnesium.[59]

Other cane nutrition problems identified by plantation research personnel included a few fields with apparent zinc deficiency, possibly caused by low ratios of zinc to phosphorus in the soil. In addition, accumulation of hexazinone (Velpar), dinitroanaline (Trifluralin and Treflan), and phenoxy (2,4-D) herbicides were suspected but were not proved to cause poor cane growth in some fields. Other problems included excessive rat damage; diseases like *Pythium* root rot, yellow leaf syndrome and red rot; and insect damage by borers and bud moths.[60]

Finally, plantation research staff noted problems in managing the new variety H 78–4153. While it had demonstrated good growth and sugar yields, it had soft stems that tended to develop cracks. The soft stems were a welcome characteristic in the factory because milling required less power. But the cracks provided a route for disease organisms to enter the stalk, reducing juice quality and sugar yields per acre. This variety also produced lush top growth when well fertilized with nitrogen and given adequate irrigation. If ripening was not carefully managed by limiting nitrogen fertilizer and reducing irrigation early enough, its lush top growth led to oversensitivity to the chemical ripener glyphosate, causing whole stalks to die prior to harvest and reducing juice quality.[61]

This report by HC&S agricultural research staff gives us an excellent snapshot of the multiple issues involved in managing a large and complex plantation like HC&S. Fortunately, attention to management issues and increased rainfall beginning in 1997 caused plantation yields to rebound in 1998 and 1999 (see Table 5.1).

Despite surviving the early 1980s and producing record crops in the mid- and late-1980s, HC&S, like the remaining Hawaiʻi sugar companies, felt that it was under attack from all sides. Declining sugar yields in the 1990s were a major problem. In addition, sugar refiners and the processed food industry were pressuring Congress to drop support for the domestic sugar program. Other sweeteners, including both corn syrups and noncaloric sweeteners, had reduced sugar's share of the sweetener market. Environmental concerns about smoke from the Puʻunēnē factory and pre-harvest cane burning were pressuring the company to investigate green cane harvesting and make factory improvements. Labor shortages on Maui had forced HC&S

to expand worker recruiting, retention, and on-the-job training programs, including the enrollment of 263 employees in the HSPA Trades Progression Program. Nevertheless, as of 1989, the company—with a total workforce of 1,250—had job vacancies in 61 categories, putting additional pressure on management to mechanize and for many employees to work overtime. In response, the company had begun to computerize its accounting, engineering design, personnel safety, affirmative action, and employee benefit records. The company's efforts did not go unnoticed, and in 1989 it received the U.S. Senate Productivity Award for large businesses. Still, despite the fact that in the 1990s the Pāʻia factory was one of the most automated in the world, in an effort to save costs, it was closed in 2000, and all milling was moved to the Puʻunēnē factory.[62]

Cane Fires and Factory Smoke

Public concerns about smoke from the HC&S factories in Pāʻia and Puʻunēnē and pre-harvest burning of cane fields began to increase in the 1980s. In an effort to maintain good public relations, HC&S installed expensive wet scrubbers at the Pāʻia factory in 1985 and at Puʻunēnē in 1987 and 1988. These scrubbers used water sprays to remove virtually all particulates (dirt and fly ash) from boiler flue gas, leaving only water vapor to form the familiar white cloud emitted from its stacks. The public also complained that drift from aerial spraying could affect nearby residents and businesses, so the plantation eliminated aerial spraying of herbicides. In response to concerns that pre-harvest burning of cane fields aggravated respiratory illnesses, the company eliminated the burning of trash in seed fields, limited pre-harvest burns to two fields per day, and distributed flyers to warn neighbors when burns would occur in their vicinity. HC&S management explained that despite extensive experimentation in the 1960s, no economical method of harvesting unburned two-year-old cane had been discovered. In addition, changing to a one-year cycle would require massive operational changes that would likely bankrupt the company. However, in an effort to reduce off-site effects of smoke, HC&S automated its 10 weather stations and began to use weather information to schedule pre-harvest cane burning to minimize smoke drifting into nearby residential areas. It also set up a system to advise its neighbors when burning of specific fields was planned in order that those sensitive to the smoke could take necessary precautions. The company was pleased when in the early 1990s a study by the National Institute of Occupational Safety and Health found no evidence of impaired lung function in HC&S workers on harvesting crews that were exposed to cane smoke.[63]

In Hawai'i, regulation of agricultural burning was the responsibility of the state's Department of Health Clean Air Branch, which issued permits annually and assessed fees based on acreage to be burned, usually about 15,000 acres per year at HC&S. The agricultural burning permits for HC&S generally restricted burning to between 6:00 a.m. and 6:00 p.m., though some fields could be burned as early as 3:00 a.m. and as late as 8:00 p.m. to take advantage of prevailing winds during those periods. The HC&S permit did not allow burning of noncrop materials and prohibited burning during "no-burn" periods designated by the Department of Health (usually declared when the normal trade winds died, preventing normal dispersal of the smoke). In response to public health concerns, monitors of particulates in smoke were set up at Kīhei in 1996 to measure the amount of smoke reaching the town. These monitors measured "inhalable coarse particles" (PM_{10}) that are usually associated with dust and "fine particles" ($PM_{2.5}$) associated with smoke and combustion gases. The U.S. Department of Agriculture reviewed the particulate monitoring data from Kīhei for 2000–2004 and found that neither the daily nor the annual PM_{10} or $PM_{2.5}$ concentrations reached the ambient air quality standards established by the U.S. Environmental Protection Agency, even for sensitive infants, the elderly, and asthmatics.[64]

Even so, the issue of pre-harvest burning of cane fields resurfaced in 2012. A new anti-burning group formed on Maui. The organization cited studies documenting the health hazards of particulates, other smoke constituents, and cane burning in Louisiana and Brazil. The effort resulted in a news article documenting the concerns of Maui residents and physicians that smoke from pre-harvest cane fires might be causing significant respiratory problems, especially in susceptible children, the elderly, and asthmatics. The article also suggested that burning the polyethylene drip tubes and polyvinyl chloride pipes used in drip irrigation systems could release dioxins that are likely carcinogens and endocrine disrupters. It cited evidence that trade winds do not disperse the smoke because the Maui volcanos create a "closed circulation system called the 'Maui Vortex' over the central valley that traps pollutants." It asserted that the single air monitor in Kīhei was not measuring toxins that had been identified in smoke produced by burning sugarcane in Florida, and it quoted sources who accused sugar industry sympathizers of threatening those who spoke out against burning.[65]

HC&S management responded to the news article by describing company efforts to reduce the impacts of burning, noting that the company had ceased to burn seed fields in the 1980s, and it used several means to notify residents of planned burns. It also worked with the Department of Health to burn only at times when weather conditions favored the rise and dispersal of the smoke plume, and it refrained from burning during "no-burn" periods. It emphasized the company's investments in factory and harvesting equipment to find alternatives to burning, including using cane trash as fuel for electricity generation and as raw material for value-added co-products such as fiberboard and cellulosic ethanol. However, the cost of hauling and processing the cane trash was higher than the value of the additional energy that could be produced from burning it. The company warned that preventing cane burning without a cost-effective alternative would jeopardize the livelihoods of 800 HC&S employees and other Maui residents that depended on the company. Not satisfied by HC&S management's response, burning opponents delivered a petition with 8,730 signatures to the Department of Health and HC&S; they also scheduled a public protest of the practice in Kahului, which was countered with a demonstration by supporters of the company. Considering the long-standing determination of burning opponents, the controversy will likely continue until HC&S either ceases to harvest cane or it finds a cost-effective technology to harvest and process unburned cane.[66]

Protecting Watersheds, Aquifers, and Agricultural Lands

Pre-harvest cane burning was not the only environmental concern for HC&S. Its success had always depended on adequate surface water, adequate groundwater, and good agricultural land, and it had a track record of supporting and often joining with other organizations to protect those resources.

The East Maui watershed, with its abundant rainfall and many streams and waterfalls, had changed dramatically since the eighteenth century. The native rainforests had been logged for iliahʻi (sandalwood), ʻōhiʻa, and koa, and numerous invasive species of plants and animals had taken up residence, often causing major disturbance due to grazing, foraging, and competition with native species. In addition, the water collection tunnels and ditches built by the sugar industry in the late-nineteenth and early-twentieth centuries had captured and diverted water from streams that had once made their way unimpeded to the sea. By the 1980s concerns were being raised that HC&S was diverting too much water from East Maui, to the detriment of both animals dependent on stream flows and farmers hoping to expand small-scale kalo (taro) production near the coast. In 1991 East Maui

Watershed Partnership was formed with the objective of pooling the economic, technical, and human resources of stakeholder organizations to protect the 120,000-acre East Maui watershed. The partnership, composed of East Maui Irrigation Company, the Nature Conservancy, Haleakalā National Park, Haleakalā Ranch Company, Hāna Ranch Partners, the County of Maui Department of Water Supply, the State of Hawai'i Department of Land and Natural Resources, and several supporting organizations, has attempted to coordinate efforts to combat invasive species and protect the watershed.[67]

The basal aquifers under Central and West Maui are the principal sources of the island's drinking water, as well as an important source of irrigation water for HC&S. The thickness of these basal aquifers is determined by the balance between recharge and discharge. During the 1990s concerns arose that the transition zone between freshwater and saltwater in the 'Iao basal aquifer was rising, raising concerns about future groundwater supplies in Central and West Maui. In response, Maui County contracted with the U.S. Geological Survey (USGS) to study the effects of changing land use, irrigation methods, and drought on groundwater recharge in West and Central Maui. Discharges from the western and central Maui aquifers include spring flows to streams from high-level aquifers, diffuse seepage to the ocean from basal aquifers, and pumping for both municipal and agricultural uses. High-level aquifers were found to contribute about 101 mgd to springs in the 14 largest gauged stream systems on West Maui. This was somewhat greater than the total discharges due to pumping, which in 2006 included about 68 mgd pumped for sugarcane irrigation by HC&S, 23 mgd for public water supply in the Wailuku area (having increased from about 10 mgd in 1970), and about 4 mgd for water supply in the Lahaina area.[68]

The decrease since 1970 in recharge of the basal aquifers under Central and East Maui by deep percolation of sugarcane irrigation water is attributed primarily to the replacement of inefficient furrow irrigation at HC&S and Wailuku Sugar Companies with much more efficient drip irrigation. In addition, Wailuku Sugar Company closed in 1988, reducing the total irrigated acreage in Central Maui. Attempting to quantify the major factors affecting the basal aquifers under Central and West Maui, USGS estimated that between the two periods of 1926–1979 and 2000–2004, average rainfall had decreased from 897 mgd to 796 mgd, irrigation had decreased from 437 mgd to 237 mgd, and groundwater recharge had decreased from 693 mgd to 391 mgd. In response, water levels in the basal aquifers under Central and West Maui declined, the transition zone between freshwater and seawater became shallower, and the chloride concentrations of wells in the area increased. Studies suggested that these trends could be minimized or reversed by better distributing well fields across the landscape and restoring stream flows to increase recharge of the ʻĪao and Waiheʻe aquifer systems. But eliminating irrigated agriculture in Central or West Maui would exacerbate the problem by reducing groundwater recharge by approximately 18 percent. Even more serious reductions in recharge would occur if a drought like that from 1998–2002 coincided with elimination of agriculture.[69]

As it moved into the second decade of the twenty-first century, HC&S owned about 43,300 acres on Maui, with 35,500 acres used for sugarcane production. Recognizing the need to protect its agricultural economy, Hawaiʻi passed legislation to safeguard agricultural lands, promote diversified agriculture, and increase the state's agricultural self-sufficiency. In response, A&B solicited protection for key HC&S lands, and in 2009 the state designated 27,000 acres of HC&S land on Maui as Important Agricultural Lands.[70]

Stream Flows

A&B owns 16,000 acres of the East Maui watershed and for many years held four licenses from the State of Hawaiʻi to use water from an additional 30,000 acres of state lands on East Maui. These licenses were used by East Maui Irrigation (EMI) to collect and transport water from East Maui watersheds for irrigation of HC&S lands. The last of the water licenses expired in 1986, and since that time all four licenses have been extended as revocable permits that are renewed annually. In 2000 EMI had 18 employees who maintained 50 miles of tunnels, 24 miles of open ditches, 355 stream diversions, 16 steel siphons, 7 storage reservoirs with a capacity of 274 million gallons, and 62 miles of roads. The many ditches built between the 1870s and 1920s had been integrated into four main transmission ditches that brought water from the East Maui watershed to the plantation. They were: Wailoa Ditch (195 mgd capacity), New Hāmākua Ditch (100 mgd capacity), Lowrie Ditch (70 mgd capacity), and ʻHaikū Ditch (70 mgd capacity). Though the system's total capacity was 435 mgd, its average delivery was 165 mgd.[71]

After Wailuku Sugar Company closed in 1988, its irrigation system, which included a number of small ditches, was consolidated into the West Maui Irrigation System. Its two ditches, Waiheʻe Ditch (70 mgd capacity) and Spreckels Ditch (60 mgd capacity), took water from seven surface water diversions draining a watershed of 13,500 acres and had a total length of 17.3

miles. Jointly operated and maintained by Wailuku Agribusiness Co. and HC&S, the system delivered an average of 45 mgd. These two ditches have been known by several confusing names. For example, the Waiheʻe Ditch, built in 1905–1907, was formerly known as Waiheʻe Canal; and Spreckels Ditch (which should not be confused with another Spreckels Ditch that is part of the East Maui Irrigation System) was formerly called the Waiheʻe Ditch.[72]

A&B's East and West Maui ditch systems were, of course, designed to intercept runoff and subsurface flows from streams and high-level aquifers and divert them to sugarcane fields in Central Maui. The original collection systems had been built in the late nineteenth and early twentieth centuries and included sections of leaky unlined ditches and tunnels that lost a significant amount of water, replenishing the streams and aquifers. However, over the decades since their construction, the ditches and tunnels were lined and otherwise improved to the point that relatively little water leaked from the system and flowed to the sea, especially during the dry summer season. The reduced stream flows at lower elevations had long been blamed for limiting traditional kalo cultivation and damaging the environment on the lower windward slopes of East Maui, as well as along four major streams draining the northeastern slopes of West Maui. These four streams—Waiheʻe, Waiehu, ʻĪao, and Waikapū—are known as Nā Wai ʻEhā or "The Four Great Waters."

Legal battles over diversion of water by the sugar industry can be traced back to the Hawaiʻi Supreme Court's decision in a case regarding water in the Waiāhole Ditch on Oʻahu. That ditch had been built between 1913 and 1916 and brought water from the windward to the leeward side of Oʻahu for sugarcane irrigation. The ditch had a normal flow of 27 mgd, but when Oʻahu Sugar Company closed in 1995, numerous entities applied for permits to use the ditch's water. After hearings and studies throughout the mid-1990s, the Hawaiʻi Commission on Water Resource Management ruled that about 14 mgd should be reserved for agricultural uses, about 7 mgd should be considered an "agricultural reserve" and "non-permitted groundwater buffer," and 6 mgd should be released to provide flow for the windward streams. Several parties appealed the commission decision to the Hawaiʻi Supreme Court, which ruled that the state's flowing streams are a public trust that should be regulated for maximum beneficial use, with provision for traditional Hawaiian rights, wildlife needs, and maintenance of ecological balance and scenic beauty. It also reversed the commission's decision to award permits to leeward agricultural interests, allocated 2.1 mgd for "system losses," and suggested that 6 mgd was too little flow to reserve for windward stream flows.[73]

In response to the Waiāhole Ditch decision, in 2001 the Native Hawaiian Legal Corporation filed a petition with the commission to amend the interim in-stream flow standards for 27 East Maui streams. Also in 2001 A&B requested the commission replace its revocable permits to East Maui water with a long-term lease. In response to these requests, the USGS conducted two studies of the effects of irrigation's water diversions on stream flows and habitat availability, which were released in 2005 and 2006. The commission also received stream survey reports from the Hawai'i Division of Aquatic Resources, held public meetings and site visits to gather facts related to the petition, and received recommendations from commission staff for new interim standards. Following many of the staff's recommendations designed to provide connectivity for aquatic life from *mauka* to *makai,* in 2008 the commissioners approved amended standards for five East Maui streams, mandating restoration to the streams of over 12.21 mgd. In 2010 the commission approved amended flow standards for six additional East Maui streams, increasing flows in two streams throughout the year and in four others only in the wet winter months. Total flow increases mandated by the new standards were 9.26 mgd from November to April and 1.1 mgd from May to October. The kalo farmers were not happy with the decision. Claiming that HC&S was wasting 66 mgd due to leaks and poor management, they requested binding arbitration in the form of a contested case proceeding. The request, initially denied by the commission, was upheld in 2012 by the State Supreme Court, but legal wrangling over stream flow standards could continue for years.[74]

Litigation regarding West Maui's Nā Wai 'Ehā waters began in 2004. Maui community groups Hui o Nā Wai 'Ehā and Maui Tomorrow Foundation petitioned the Hawai'i Commission on Water Resources to establish stream flow standards for four West Maui streams (Waihe'e, Waiehu, 'Īao, and Waikapū) and allocate a major fraction of their flow being diverted by Wailuku Water Company (formerly Wailuku Sugar Company) and HC&S to kalo farmers and the environment. The Nā Wai 'Ehā surface water management area was established in 2008 to address these concerns. In response, HC&S, citing over 100 years of historical use for sugarcane irrigation from Nā Wai 'Ehā waters, applied for two water use permits to irrigate two groups of fields containing a total of 5,441 acres with 46.87 mgd (7,098 gallons per acre per day). The requested irrigation amounts were based on the company's actual irrigation amounts for these field areas in previous years.[75]

The commission asked the State Office of Hawaiian Affairs for comment, and it objected to the request for several reasons. It stated that no request for water should be considered until interim in-stream flow standards had been established for the four streams. In addition, Hawaiian Affairs argued that HC&S had overestimated the amount of cane acreage in the areas, overestimated how much water was needed to adequately irrigate the cane, failed to show that it could not feasibly reduce its system losses, and failed to demonstrate that it could not reasonably and economically make use of existing wells to irrigate some of the fields.[76]

When in 2010 the commission set interim in-stream flow standards for Nā Wai ʻEhā waters, the petitioners appealed the decision to the State Supreme Court, arguing that the standards did not allocate sufficient water to the environment and kalo farmers. The Supreme Court ruled that, based on the public trust doctrine, the commission should reconsider its decision. In response, the commission ruled in 2012 that Wailuku Water Company and HC&S should divert 12.5 mgd of Spreckels Ditch's normal 70 mgd flow to Nā Wai ʻEhā streams: 10 mgd to Waiheʻee Stream, 1.6 mgd to North Waiehu Stream, and 0.9 mgd to South Waiehu Stream. No additional water was allocated to ʻĪao or Waikapū Streams.[77]

Prior to the commission's rulings, HC&S was using an average of 167 mgd from East Maui streams and 70 mgd from West Maui streams. In addition, it was pumping an average of 72 mgd from 16 Central Maui wells that were becoming more brackish as a result of a three-year drought and the highly efficient drip irrigation systems it had installed in the 1970s and 1980s. The loss of approximately 34 mgd in the wet season and 28 mgd in the dry season from the average 237 mgd it received from East and West Maui ditch flows were a major concern for HC&S management, especially since A&B Agribusiness was trying to recover from $45 million in operating losses in 2008 and 2009, due in large part to the ongoing drought.[78]

In response to the commission's revision of several in-stream flow standards of East Maui streams in 2008 and 2010, in 2011 the USGS cooperated with East Maui Irrigation to quantify seepage losses and gains in the four primary ditches (Wailoa, New Hāmākua, Lowrie, and ʻHaikū Ditches), as well as smaller ditches that connect with them (Koʻolau, Spreckels, Kaʻūhikoa, Spreckels at Pāpaʻaʻea, Manuel Luis, and Center Ditches). The study could measure losses and gains for only 26 reaches with a total length of about 15 miles, representing 23 percent of the total length of ditches in the EMI system. In addition, measurements were made only during periods of stable ditch flow in June, August, and September 2011. Nevertheless, the limited measurements suggested that Koʻolau and Spreckels at Pāpaʻaʻea ditches generally had seepage losses, while Wailoa, Kaʻūhikoa, and New Hāmākua ditches had seepage gains. The Manuel Luis, Center, Lowrie, and ʻHaikū ditches had both losses and gains. Open ditches had seepage losses that ranged from 0.1 cubic feet per second (cfs) (0.06 mgd) to 3.0 cfs (1.9 mgd) per mile of ditch. Tunnels generally had seepage gains that ranged from 0.1 cfs (0.06 mgd) to 5.2 cfs (3.4 mgd) per mile. Expressed as a percentage of flow, seepage losses were greater at low than at high flows. While these losses at times of water shortage are detrimental to HC&S, they benefit vegetation and stream flows below the ditches. Because the study represented only 23 percent of the EMI system and both gains and losses depend on weather and flow rates, a much more detailed study would be required to provide credible estimates of average annual seepage gains and losses from the system.[79]

Sugar Prices

The 2002 and 2008 Farm Bills used two mechanisms—a tariff-rate quota import system and a price support loan program—to assure a reliable domestic supply of sugar at reasonable prices. The import quota system limited imports from countries other than Mexico and Canada by allowing only limited imports with relatively low tariffs. Above these quota amounts, higher over-quota tariffs were imposed. Since most of C&H's refined sugar was sold west of the Mississippi River, the region that also produces a large portion of U.S. beet sugar, HC&S's main domestic competition was beet sugar produced in the western United States. The domestic price support program set an effective floor on domestic prices, providing sugar-secured loans to producers at a loan rate, or "support price," of 18 cents per pound for 2003–2007. The support price rose to 18.5 cents per pound for 2010–2011 and 18.75 cents per pound for 2012–2013. In addition, if the market was "oversupplied," a limited amount of sugar could be diverted to ethanol production. Prior to 2009, the sugar programs of successive farm bills had caused domestic raw sugar prices to remain relatively stable. With the exception of a period of relatively low prices in 1999–2000, prices almost always remained at 20–24 cents per pound from 1996 through 2008. However, beginning in 2009, U.S. raw sugar prices increased dramatically due to tight world supplies, fluctuating from 28–42 cents per pound from 2009–2011, though they began to decline in 2012.[80]

HC&S in the Twenty-First Century

As HC&S moved into the new century, it faced increased production, economic, and environmental challenges. A major change occurred in September 2000, when the Pāʻia mill closed, and all the cane on the plantation's 35,000 acres began to be milled and processed at the Puʻunēnē mill.[81] The company continued to produce raw sugar, selling about 90 percent of its production to C&H Sugar Company for refining and sale, primarily in the western and central United States. This sugar was transported to the C&H refinery in California by the Hawaiian Sugar & Transportation Cooperative, a sugar grower cooperative that, due to the closure of all other sugar companies in the state, had only one member: HC&S. The cooperative sold its raw sugar to C&H at the New York No. 16 Contract settlement price, less a volume discount. The remaining 10 percent of HC&S sugar was sold as specialty food-grade sugars. As in past decades, HC&S sold a fraction of the electric power it generated from burning bagasse and from hydropower. Maui Electric Company purchased renewable energy from HC&S at a price equal to the utility's "avoided cost" of not producing the power itself using imported oil. In 2013 HC&S generated about 194,200 MWh and sold about 58,900 MWh to Maui Electric Company.[82]

Thanks to crop management changes begun in 1997, HC&S maintained sugar yields above 12 tons per acre from 1998–2003, with three of those years producing yields above 13 tons per acre. However, much as they had in the mid-1990s, sugar yields began to fall in about 2004, dipping below 11 tons per acre in 2006 then falling below 10 tons per acre in 2007 and below 9 tons per acre in 2008–2009. They finally began to recover in 2010. The importance of adequate irrigation water to sugar production is clear from Table 5.2. The three years with the lowest yields in the decade 2003–2012 were 2008 and 2009. Since sugarcane is a two-year crop at HC&S, both these crop suffered through 2008, the driest year in over 80 years of rainfall and ditch flow records kept by HC&S. Total water available was only 62.6 billion gallons, far less than the average of 79.4 billion gallons for the decade. In addition, the 2007 crop was reduced by a less severe dry period in 2006, and the 2007, 2008, and 2009 crops produced the lowest total and per-acre sugar yields in decades. As a result of the dry weather conditions, in 2008 HC&S furloughed hundreds of workers, beginning in the summer when seed cane cutting, field preparation, and planting were temporarily suspended due to lack of water needed to germinate the seed pieces.

Table 5.2. Annual water supplies received by HC&S from East Maui and West Maui ditches and wells, 2003–2012

Year	East Maui	West Maui	Wells	Total	Sugar	Sugar yield
	Million gallons				1000 tons/acre	
2003	48,159	9,903	23,850	81,912	206	13.1
2004	52,755	12,474	15,264	80,493	199	11.8
2005	61,905	14,638	15,850	92,393	193	11.6
2006	48,776	11,327	14,812	74,914	174	10.2
2007	47,235	11,525	23,254	82,014	165	9.7
2008	29,858	9,998	22,716	62,572	145	8.6
2009	49,528	11,803	20,672	82,003	127	8.4
2010	41,408	10,032	13,956	65,396	171	11.1
2011	42,324	8,077	32,728	83,130	183	12.1
2012	50,219	9,514	29,895	89,628	178	—

Source: Alexander & Baldwin, Financial reports.

In December the company implemented a one-week phased furlough, affecting 674 workers at all levels of the organization.[83]

In part because of declining yields, operating profits of A&B's agribusiness component, largely composed of HC&S, fell from an average of $7.9 million for 2001–2006 to $0.2 million in 2007. They then plummeted to operating losses of $12.9 million in 2008 and $27.8 million in 2009. HC&S responded by drastically cutting capital investments in 2009–2010. Operating profits rebounded in 2010–2013 due to high domestic prices and the return of more normal ditch flows and sugar yields, growing to $6.1 million in 2010, $22.2 million in 2011, $20.8 million in 2012, and 10.7 million in 2013 (see Tables 5.2 and 5.3).[84]

In addition to its production challenges, HC&S continued to face, as it had for decades, uncertainty about world sugar prices and the domestic sugar price support system. It also remained a target of public concerns for maintaining good air quality, protecting "mauka to makai" stream flows throughout the year, and providing water for diversified agriculture.

Exploring options to diversify, HC&S leased several hundred acres of land to seed companies, primarily for corn seed production. The success of sugarcane ethanol production in Brazil and the potential for producing electricity and liquid fuels from sugarcane biomass raised hopes of a new and more profitable business model. In response, HC&S conducted research on harvesting two-year-old unburned cane with large combine harvesters and sought assistance from the U.S. government to diversify into greater

Table 5.3. Selected production and economic statistics for HC&S and A&B Agribusiness—2001–2013

	2001	2002	2003	2004	2005	2006	2007	2008	2009	2010	2011	2112	2013
Operating profit ($ millions)	5.7	13.8	5.1	4.8	11.2	6.9	0.2	(12.9)	(27.8)	6.1	22.2	20.8	10.7
Harvested (1,000 acres)	15.1	16.6	15.6	16.9	16.6	17.0	16.9	17.0	15.0	15.5	15.1	15.9	15.4
Sugar produced (1,000 tons)	192	216	206	199	193	174	165	145	127	171	183	178	192
Sugar yield (tons/acre)	12.7	13.0	13.1	11.8	11.6	10.2	9.7	8.6	8.4	11.1	12.1	11.3	12.4
Total water avail. (billion gallons)	—	—	81.9	80.5	92.4	74.9	82.0	62.6	82.0	65.4	83.1	—	—
Molasses (1,000 tons)	71.2	74.3	72.5	65.1	57.1	55.9	51.7	52.2	41.7	52.8	53.1	50.5	54.8
Capital invest. ($ millions)	9.5	9.9	12.6	10.2	13.0	15.0	20.5	15.2	3.4	6.8	10.5	11.7	11.6
Energy produced (GWh)	204	220	212	209	219	208	218	211	188	190	191	182	194
Energy sold (GWh)	61	87	82	94	96	98	94	91	73	68	65	58	59

Source: Alexander & Baldwin, Financial reports.

production of renewable energy. In 2010 the U.S. Department of Energy and the Office of Naval Research each began providing $2 million annually to help HC&S achieve its goal of becoming a large-scale biomass energy farm. The DOE funding was directed to the University of Hawai'i for research on energy crop development and technologies to convert biomass to liquid fuels. The Office of Naval Research also supported research by the U.S. Department of Agriculture to evaluate alternative energy crops, assess long-term water resource requirements for biomass production, and develop computer models to manage its irrigation more effectively. [85]

Despite two very profitable years in 2011 and 2012 owing to high sugar prices and good yields, A&B management was concerned that the next Farm Bill would not maintain adequate sugar price supports. They also remained concerned about recent and possible future losses of their rights to use irrigation water from the East and West Maui watersheds. A&B's 2011 Annual Report stated that the East Maui and West Maui irrigation systems provided 58 percent and 14 percent, respectively, of the irrigation water used by HC&S over the previous decade. Losses imposed by the Hawai'i Commission on Water Resource Management in response to litigation, while not threatening near-term sugar production, "will result in a future suppression of sugar yields and will have an impact on the Company that will only be quantifiable over time."[86]

So, how far ahead can we look to predict the future of HC&S and A&B's agribusiness operations? Probably not far. Threats identified by A&B include:[87]

- lack of irrigation water as a result of either lack of rainfall in the East and West Maui watersheds or litigation that limits access to runoff;
- declining sugar prices, whether influenced by competition with domestic sweeteners or changes in U.S. sugar support or trade policies; and
- HC&S's ability to produce raw sugar economically, which could be affected by weather, natural disasters, sugarcane diseases, weed control, uncontrolled fires (including arson), regulations restricting pre-harvest cane burning, increases in input costs (fuel, fertilizer, herbicide, drip tubing), major equipment failures in the factory or power plant, and labor availability and cost.

Uncertainty regarding government support of sugar prices increases as each new Farm Bill is negotiated, usually every four to six years. But the 2014 Farm Bill left sugar programs unchanged, providing the industry with continuing sugar price supports. World prices can change weekly, though, and major adjustments can occur in less than a year, drastically affecting company profitability from one year to the next. For example, A&B Agribusiness, composed primarily of HC&S, was quite profitable from 2010 through 2013 (see Table 5.3), due in large part to high world sugar prices. However, world prices declined from 20–24 cents per pound in 2012 to 16–20 cents per pound in 2013, comparable to the cane sugar support price of 18.75 cents per pound. Despite this drop in prices, A&B Agribusiness still posted an operating profit of $10.7 million (over 7 percent profit) in 2013.

In view of recent litigation and public concern over water use and the health effects of cane burning, HC&S can anticipate continued legal and public relations battles. Loss of a fraction of East or West Maui irrigation rights would likely reduce HC&S's sugar production unless means are found to reduce irrigation system losses, which could require very costly infrastructure improvements. As they have periodically throughout the history of the industry, droughts could reduce rainfall on the East Maui watershed while increasing temperatures and irrigation demands in sunny Central Maui. Finally, regulations to greatly reduce pre-harvest burning would substantially increase the cost of harvesting, transporting, and milling the cane, as well as reduce the percentage of sugar in the cane that could be extracted and sold.

The Hawai'i sugar industry has long looked to product diversification to both increase and stabilize its profitability. Before World War II the industry produced much of the food consumed by its workforce and provided much of the electricity used in the neighbor islands. The founders of the industry expanded into pineapple and cattle ranching in the nineteenth century, and the plantations on all the islands maintained beef and dairy herds for decades. Rice, coffee, pineapples, bananas, citrus, potatoes, macadamia nuts, rum, flowers, and field crop seeds have all, at one time or another, been considered growth industries in Hawai'i. In addition, the sugar industry has long hoped to produce building materials and industrial chemicals from by-products like bagasse, demonstrating the ability to produce a variety of value-added products but never achieving long-term commercial viability except for bagasse-based wall boards.

With this background, which future agricultural business models might make sense for A&B and HC&S? Of course, if the threats described above do not make sugar and electric energy production unprofitable, the plantation could continue to operate as it has in recent years, improving wherever possible its production and labor efficiencies. Possible cane production alternatives could include changing from recumbent two-year crops to standing one-year crops, then using combines to harvest the standing cane—like those used in many other sugarcane industries worldwide. This would probably require the selection of new upright sugarcane varieties, redesign of fields, rock removal, conversion from surface to subsurface drip irrigation, and the harvesting of several ratoon crops before replanting. Of course, the Experiment Station and plantations have experimented for decades with equipment needed to produce one-year cane; however, considering the large costs of conversion, no economically viable system has yet been identified.

Many have felt that the industry's future lies in some combination of sugar and renewable energy production.[88] This could include green harvest of at least some of its acreage and using the extra biomass contained in unburned tops to increase steam and electricity production. All or some of the molasses produced, as well as some of the raw cane juice, could be fermented and distilled into ethanol. A flexible system like that used in the Brazilian sugar industry would allow HC&S to vary the percentage of cane used for ethanol and raw sugar production, depending on the relative prices of the two commodities. Another possibility might be to convert part of the bagasse to syngas, then reform it catalytically to produce ethanol or diesel; however, such a system remains to be demonstrated at a commercial scale. The major issues with these alternatives are (1) the capital costs of building the distillery and/or syngas transformation facility; (2) the economic and engineering feasibility of harvesting, transporting, and grinding green cane; and (3) the environmental challenges and costs of treating and disposing of large amounts of vinasse, the liquid waste of the fermentation and distillation processes.

If sugar price supports were eliminated or reduced for an extended period, A&B could decide to close HC&S and convert its 35,000 acres of cane land to other uses. For many years commercial, industrial, and residential development have nibbled away at sugarcane lands across Hawai'i. In addition, Maui has recent experience with the economic impacts of closing Wailuku Sugar Company in 1988 and Pioneer Mill Company in 1999. Of course, neither of these plantations approached the size of HC&S, and many of their fields near the coast were ideal for real estate development. That is definitely not the case for much of HC&S's massive acreage. If its fields and irrigation systems were simply abandoned to revert to dry shrub and grasslands, Maui residents and tourists would certainly object. Therefore, it would likely be necessary for A&B to find other irrigated agricultural uses for many of its cane fields.

If A&B decided to become a diversified agriculture company, it might choose to produce crops and/or livestock that already have significant markets and production in the state. According to the USDA, Hawai'i's top categories of agricultural products in 2011 were seed, mostly hybrid corn ($247 million), fruits and nuts ($100 million), livestock products ($90 million), cane for sugar ($78 million), vegetables and melons ($82 million), and greenhouse and nursery crops ($72 million). However, in 2007 only about 100,000 acres of all crops were harvested in the entire state. It was clear that the markets for these crops would not support the additional production possible on up to 35,000 acres of irrigated HC&S lands and that export of products would be required.[89]

Over the past several decades, agricultural seed production has grown into the state's largest agricultural industry, producing $247 million worth of seed in 2011, almost tripling the value of sugar produced by HC&S, and making Hawai'i the third-ranking state in seed exports.[90] Since HC&S has already leased land to the seed industry, expansion of seed production might allow the company to gradually move more of its land away from sugar production. However, seed production is a mature and complex business requiring extensive expertise in plant breeding, biotechnology, and marketing. As a result, HC&S might choose to partner with or sell or lease its land to one or more major seed companies.

Summary

U. S. government sugar programs maintained relatively stable sugar prices from the 1930s through the mid-1970s, enabling Hawai'i's sugar industry to remain profitable. However, a dramatic but short-lived increase in world market prices in 1974–1975 was followed by elimination of federal sugar price supports, increased world production, and a collapse of world prices. Though a federal sugar program was reinstated in 1977, prices remained low, and 12 sugar companies closed between 1970 and 1980. The remaining 16 plantations tightened their belts and reduced costs of production. On irrigated plantations like HC&S, the most important factor in reducing costs of production was conversion from furrow to drip irrigation in the late 1970s and early 1980s. This massive transformation of farming practices was made possible by Experiment Station, HSPA research in cooperation with plantations, and it allowed several plantations to remain in business by increasing yields and reducing labor costs.

By the late 1980s it was clear that the industry had to diversify in order to survive. In response, the Experiment Station broadened its research portfolio to include a number of tropical crops for food and energy, as well as industrial chemicals and products that could be made from sugarcane and sugar. Though it demonstrated success in producing a variety of crops and products, research failed to identify adequate markets for profitable production. Though domestic sugar prices stabilized somewhat, fluctuating between 18 and 24 cents per pound from 1982 through 2009, inflation continued to eat away at profits, and increasing use of corn syrups and noncaloric sweeteners dampened demand for sucrose. Companies continued to close, and in 1996, after 100 years of voluntary support by the sugar industry, the Experiment Station was reorganized and renamed the Hawaii Agriculture Research Center (HARC). By 1997 only five sugar companies remained, three on Kaua'i and Pioneer Mill Company and HC&S on Maui. But Pioneer Mill Company closed in 1999, and the three plantations on Kaua'i closed in 2000 and 2010. As of 2011 only HC&S remained.

During recent years HC&S has struggled to maintain its historical water rights in the face of legal challenges seeking to restore stream flows and provide agricultural water to others. It has also worked with surrounding communities to reduce exposure to smoke from pre-harvest cane fires. It has leased land and water to Hawai'i's burgeoning seed production industry, and it has continued to conduct research on biomass energy crops and processes needed to increase its production and sale of electricity. It has also considered production of oil crops for biodiesel and sugarcane for ethanol and other liquid fuels.

After suffering through below-average yields in 2007–2009 due largely to drought, HC&S rebounded with increasing yields in 2010–2013. This, along with increased world sugar prices, allowed the company to post substantial operating profits in 2010, 2011, 2012, and 2013. However, questions remain. Will HC&S, the only remaining sugar company in Hawai'i, be able to remain profitable in a constantly changing economic, political, and regulatory environment? And will it be able to work together with its neighbors to keep Central Maui green and productive, like the sugar industry has for over 150 years?

Notes

Chapter 1: Birth of an Industry—to 1875

1. L. D. White, *Canoe Plants of Ancient Hawai'i*. Available online at http://www .canoeplants.com/ (accessed September 22, 2011).

2. Beatrice H. Krauss, *Native Plants Used as Medicine in Hawai'i* (Honolulu: Harold L. Lyon Arboretum, University of Hawai'i at Manoa, 1979).

3. James Cook, The Three Voyages of James Cook Round the World Complete in Seven Volumes, Vol. VI, Being the Second of the Third Voyage (London: Longman, Hurst, Rees, Orme, and Brown, 1821), 179.

4. Ibid., 189.

5. Ibid., 485–486.

6. Ibid., 490–491.

7. Ibid., 486.

8. Department of Foreign Affairs, Kingdom of Hawai'i, "Coffee, the coming staple product," in *The Hawaiian Islands: Their Resources Agricultural, Commercial and Financial* (Honolulu: Hawaiian Gazette Company, 1896); J. M. Lydgate, "Sandalwood days," in *Hawaiian Almanac and Annual for 1916* (Honolulu: Thos. G. Thrum, 1915).

9. G. Daws, *Shoal of Time: A History of the Hawaiian Islands* (Honolulu: University of Hawai'i Press, 1968).

10. Archibald Campbell, *A Voyage Round the World from 1806 to 1812, in Which Japan, Kamschatka, the Aleutian Islands, and the Sandwich Islands Were Visited* (New York: Van Winkle, Wiley & Co, 1817). Available at http://books.google.com /books?id=jkoMAAAAYAAJ.

11. Campbell, *A Voyage Round the World;* Daws, *Shoal of Time;* H. A. Wadsworth, "A historical summary of irrigation in Hawai'i," *Hawaiian Planters' Record* 37, no. 3 (1933): 124–162.

12. W. D. Alexander, "A brief history of land titles in the Hawaiian Kingdom," in *Hawaiian Almanac and Annual for 1891* (Honolulu: Thos. G. Thrum, 1890), 105–124.

13. Alexander, "A brief history of land titles in the Hawaiian Kingdom."

14. Alexander, "A brief history of land titles in the Hawaiian Kingdom."

15. M. M. O'Shaughnessy, "Irrigation," *The Planters' Monthly* 21 (1902): 615–622.

16. Wadsworth, "A historical summary of irrigation in Hawai'i."

17. E. M. Nakuina, "Ancient Hawaiian water rights," in *Hawaiian Almanac and Annual for 1894* (Honolulu: Thos. G. Thrum, 1893), 79–84.

18. Wadsworth, "A historical summary of irrigation in Hawai'i."

19. Nakuina, "Ancient Hawaiian water rights."

20. Ibid.

21. Ibid.

22. Ibid.

23. Ibid.

24. Campbell, *A Voyage Round the World*, 100, 111–112.

25. Wadsworth, "A historical summary of irrigation in Hawai'i."

26. Ibid.

27. Campbell, *A Voyage Round the World*, 113–114.

28. Department of Foreign Affairs, "Coffee, the coming staple product"; Lydgate, "Sandalwood days"; Hawai'i Department of Agriculture, *History of Agriculture in Hawai'i*. Available at http://hawaii.gov/hdoa/ag-resources/history.

29. "Our forestry problems as seen by Hillebrand in 1856," *Hawaiian Planters' Record* 22 (1920): 143–144.

30. Alexander, "A brief history of land titles in the Hawaiian Kingdom."

31. Ibid.

32. Ibid.

33. Perry, Antonio. "A Brief History of Hawaiian Water Rights." Honolulu: n.p., 1912. TS. Hamilton Lib., University of Hawaii, Manoa; Wadsworth, "A historical summary of irrigation in Hawai'i."

34. Hawai'i Department of Agriculture, *History of Agriculture in Hawai'i*.

35. Department of Foreign Affairs, "Coffee, the coming staple product."

36. Thomas G. Thrum, *Hawaiian Almanac and Annual for 1877* (Honolulu: Thos. G. Thrum, 1876). (Entire works by Thrum hereafter cited as Thrum, *Hawaiian Almanac [year]*.)

37. Isabella L. Bird, *The Hawaiian Archipelago* (1875; Project Gutenberg, 2004), http://www.gutenberg.org/ebooks/6750.

38. Cook, The Three Voyages of James Cook Round the World, p. 189. D. D. Baldwin, "Lahaina cane," *The Planters' Monthly* 1 (1883): 42–43; Hawaiian Sugar Planters' Association, *Story of Sugar in Hawai'i* (Honolulu: Hawaiian Sugar Planters' Association, 1929).

39. Ronald Takaki, *Pau Hana: Plantation Life and Labor in Hawai'i, 1835–1920* (Honolulu: University of Hawai'i Press, 1983).

40. Ibid.

41. George M. Rolph, *Something about Sugar* (San Francisco: John J. Newbegin, 1917).

42. Thrum, *Hawaiian Almanac 1875*.

43. Takaki, *Pau Hana*.

44. Thrum, *Hawaiian Almanac 1875*.

45. Rolph, *Something about Sugar*.

46. Thrum, *Hawaiian Almanac 1875*.

47. Rolph, *Something about Sugar*.

48. Thrum, *Hawaiian Almanac 1875*.

49. Rolph, *Something about Sugar*; Wadsworth, "A historical summary of irrigation in Hawai'i"; Thrum, *Hawaiian Almanac 1875*.

50. Rolph, *Something about Sugar*; Hawai'i Department of Agriculture, *History of Agriculture in Hawai'i*; Wadsworth, "A historical summary of irrigation in Hawai'i"; Thrum, *Hawaiian Almanac 1875*.

51. Daws, *Shoal of Time*; H. C. Prinsen Geerligs, *The World's Sugar Cane Industry: Past and Present* (Manchester: Norman Rodger, Altrincham, 1912).

52. Bird, *The Hawaiian Archipelago*.

53. Thrum, *Hawaiian Almanac 1875*.

54. Baldwin, "Lahaina cane"; Thrum, *Hawaiian Almanac 1875*.

55. J. W. Vandercook, *King Cane: The Story of Sugar in Hawai'i* (New York: Harper & Brothers Publishers, 1939); Thrum, *Hawaiian Almanac 1875*; Thrum, *Hawaiian Almanac 1893*.

56. W. W. Goodale, "Brief history of Hawaiian unskilled labor," in *Hawaiian Almanac and Manual for 1914* (Honolulu: Thos. G. Thrum, 1913).

57. Hawaiian Sugar Planters' Association, *Story of Sugar in Hawai'i;* Vandercook, *King Cane*.

58. Hawaiian Sugar Planters' Association, *Story of Sugar in Hawai'i;* Vandercook, *King Cane;* Thrum, *Hawaiian Almanac 1894*.

59. Rolph, *Something about Sugar*.

60. Bird, *The Hawaiian Archipelago*.

61. Wadsworth, "A historical summary of irrigation in Hawai'i."

62. W. H. Dorrance, *Sugar Islands, The 165-Year Story of Sugar in Hawai'i* (Honolulu: Mutual Publishing, 2000).

63. Ibid.

64. Ibid.

65. The Planters' Labor and Supply Company of the Hawaiian Islands, *The Planters' Monthly* (1883): 147–149.

66. Ibid., 149–150.

67. Ibid., 150–151.

68. Ibid., 148–149.

69. Dorrance, *Sugar Islands*.

70. Photos on exhibit at the Alexander & Baldwin Sugar Museum, Pu'unēnē, Maui.

71. Dorrance, *Sugar Islands*.

72. Bird, *The Hawaiian Archipelago*.

73. Dorrance, *Sugar Islands*.

74. Bird, *The Hawaiian Archipelago*.

75. Dorrance, *Sugar Islands*.

76. H. E. Daniel, ed., "A&B, land and sea, one hundred and twenty-five years strong," *Ampersand*, 1995; Rolph, *Something about Sugar*; R. J. Pfeiffer, "Eighty-five years a corporation: 1900–1985," *Ampersand*, 1985.

77. Prinsen Geerligs, *The World's Sugar Cane Industry;* Goodale, "Brief history of Hawaiian unskilled labor"; Rolph, *Something about Sugar*.

78. Bird, *The Hawaiian Archipelago*.

79. Ibid.

80. Ibid.

81. Ibid.

Chapter 2: Sugar Booms—1876 to 1897

1. Thomas G. Thrum, *Hawaiian Almanac and Annual for 1877* (Honolulu: Thos. G. Thrum, 1876). (Entire works by Thrum hereafter cited as Thrum, *Hawaiian Almanac [year]*.) Thrum, *Hawaiian Almanac 1914;* W. H. Dorrance, *Sugar Islands, The 165-Year Story of Sugar in Hawai'i* (Honolulu: Mutual Publishing, 2000).

2. W. W. Goodale, "Brief history of Hawaiian unskilled labor," in *Hawaiian Almanac and Annual for 1914* (Honolulu: Thos. G. Thrum, 1913); H. C. Prinsen Geerligs, *The World's Sugar Cane Industry: Past and Present* (Manchester: Norman Rodger, Altrincham, 1912).

3. Planters' Labor and Supply Company of the Hawaiian Islands, *The Planters' Monthly*, 1883.

4. Planters' Labor and Supply, *The Planters' Monthly*, 1883.

5. Ibid., 23.

6. Ibid., 23–31.

7. G. Daws, *Shoal of Time: A History of the Hawaiian Islands* (Honolulu: University of Hawai'i Press, 1968).

8. Ibid.

9. Goodale, "Brief history of Hawaiian unskilled labor"; Thrum, *Hawaiian Almanac 1880*.

10. Planters' Labor and Supply, *The Planters' Monthly*, 1883.

11. Ibid.

12. Ibid., 240–241.

13. Planters' Labor and Supply, *The Planters' Monthly*, 1886, 224.

14. Ibid., 241.

15. Ibid., 242.

16. Thrum, *Hawaiian Almanac 1891*.

17. Department of Foreign Affairs, Kingdom of Hawai'i, "Coffee, the coming staple product," in *The Hawaiian Islands, Their Resources Agricultural, Commercial and Financial* (Honolulu: Hawaiian Gazette Company, 1896).

18. Ronald Takaki, *Pau Hana: Plantation Life and Labor in Hawaii, 1835–1920* (Honolulu: University of Hawai'i Press, 1983).

19. Planters' Labor and Supply, *The Planters' Monthly*, 1886, 239–246.

20. Ibid., 241.

21. Ibid., 243.

22. Planters' Labor and Supply, *The Planters' Monthly*, 1883, 4.

23. Ibid., 125–129.

24. Ibid., 125–129.

25. Planters' Labor and Supply, *The Planters' Monthly*, 1886, 238–239.

26. Ibid., 243–244.

27. Planters' Labor and Supply, *The Planters' Monthly*, 1883, 4.

28. Ibid., 238–239.

29. Planters' Labor and Supply, *The Planters' Monthly*, 1886, 243–244.

30. Planters' Labor and Supply, *The Planters' Monthly*, 1888, 537–543.

31. Ibid., 537–543.

32. Planters' Labor and Supply, *The Planters' Monthly*, 1886, 244.

33. Ibid., 244.

34. Planters' Labor and Supply, *The Planters' Monthly*, 1883, 134–135; Planters' Labor and Supply, *Planters' Monthly*, 1886, 244–245.

35. Planters' Labor and Supply, *The Planters' Monthly*, 1883, 129–130.

36. Ibid., 129–130.

37. Ibid., 97–98.

38. Ibid., 134.

39. Ibid., 244–245.

40. Dorrance, *Sugar Islands;* Hawai'i Department of Agriculture, *History of Agriculture in Hawai'i,* http://hawaii.gov/hdoa/ag-resources/history; R. J. Pfeiffer, "Eighty-five years a corporation: 1900–1985," *Ampersand*, 1985; H. A. Wadsworth, "A historical summary of irrigation in Hawaii," *Hawaiian Planters' Record* 37, no. 3 (1933): 124–162.

41. Thrum, *Hawaiian Almanac 1891.*

42. Ibid.

43. Carol Wilcox, *Sugar Water: Hawaii's Plantation Ditches* (Honolulu: University of Hawai'i Press, 1996).

44. Pfeiffer, "Eighty-five years."

45. Ibid.

46. Ibid.

47. Jacob Adler, *Claus Spreckels: The Sugar King in Hawaii* (Honolulu: University of Hawai'i Press, 1966); Pfeiffer, "Eighty-five years."

48. J. W. Vandercook, *King Cane: The Story of Sugar in Hawaii* (New York: Harper & Brothers Publishers, 1939).

49. M. M. O'Shaughnessy, "Irrigation," *The Planters' Monthly* 21 (1902): 615–622.

50. Larry Ikeda, "HC&S Centennial," *Ampersand*, Spring 1982.

51. Ibid.

52. Ikeda, "HC&S Centennial"; Pfeiffer, "Eighty-five years."

53. Planters' Labor and Supply, *The Planters' Monthly*, 1886, 224.

54. *Harper's Weekly*, November 10, 1888.

55. Lee Meriwether, "A plantation in Hawaii," *Harper's Weekly*, November 10, 1888.

56. Ikeda, "HC&S Centennial."

57. Ibid.

58. Pfeiffer, "Eighty-five years."

59. Planters' Labor and Supply, *The Planters' Monthly*, 1883, 154.

60. Ibid., 106–107.

61. G. W. Wilfong, "Twenty years' experience in cane culture," *The Planters' Monthly* 1 (1882): 147–152. Quoted in Wadsworth, "A historical summary of irrigation in Hawaii."

62. Wilcox, *Sugar Water.*

63. Ibid.

64. Ibid.

65. Hawai'i Department of Agriculture, *History of Agriculture in Hawai'i.*

66. Planters' Labor and Supply, *The Planters' Monthly*, 1883, 137–141.

67. Isabella L. Bird, *The Hawaiian Archipelago* (1875; Project Gutenberg, 2004), http://www.gutenberg.org/ebooks/6750.

68. Thrum, *Hawaiian Almanac 1877*; Department of Foreign Affairs, "Coffee, the coming staple product."

69. Thrum, *Hawaiian Almanac 1891*; Hawai'i Department of Agriculture, *History of Agriculture in Hawai'i;* Department of Foreign Affairs, "Coffee, the coming staple product."

70. Hawai'i Department of Agriculture, *History of Agriculture in Hawai'i;* David L. Crawford, *Hawaii's Crop Parade: A Review of Useful Products Derived from the Soil in the Hawaiian Islands, Past and Present* (Honolulu: Advertiser Publishing Company Ltd., 1937).

71. Department of Foreign Affairs, "Coffee, the coming staple product."

72. Ibid.

73. Hawai'i Department of Agriculture, *History of Agriculture in Hawai'i.*

74. A. R. Grammer, "A history of the Experiment Station of the Hawaiian Sugar Planters' Association, 1895–1945," *Hawaiian Planters' Record* 51 (1947): 177.

Chapter 3: Industry Growth and Labor Unrest—1898 to 1929

1. H. C. Prinsen Geerligs, *The World's Sugar Cane Industry: Past and Present* (Manchester: Norman Rodger, Altrincham, 1912).

2. U.S. Department of Commerce, *The Cane Sugar Industry,* Miscellaneous Series no. 53 (Washington, DC: U.S. Government Printing Office, 1917).

3. Ibid.

4. U.S. Department of Commerce, *The Cane Sugar Industry;* Museum exhibits. Alexander & Baldwin Sugar Museum, 3957 Hansen Rd, Puunene, HI 96732.

5. U.S. Department of Commerce, *The Cane Sugar Industry.*

6. Experiment Station, Hawaiian Sugar Planters' Association (HSPA), *The Relation of Applied Science to Sugar Production in Hawai'i* (Honolulu: HSPA, 1915); U.S. Department of Commerce, *The Cane Sugar Industry.*

7. U.S. Department of Commerce, *The Cane Sugar Industry.*

8. *Maui News,* May 26, 1916.

9. Experiment Station, *Applied Science;* James W. Girvin. *The Master Planter* (Honolulu: Press of Hawaiian Gazette Co., 1910); U.S. Department of Commerce, *The Cane Sugar Industry;* Prinsen Geerligs, *The World's Sugar Cane Industry.*

10. U.S. Department of Commerce, *The Cane Sugar Industry.*

11. Ibid.

12. Ibid.

13. R. M. Allen, "Information for the irrigator," *Hawaiian Planters' Record* 22 (1920): 145–164.

14. Ibid.

15. Ibid.

16. J. M. Watt, "The control of irrigation water," *Hawaiian Planters' Record* 30 (1926): 195–201.

17. H. A. Wadsworth, "A historical summary of irrigation in Hawai'i," *Hawaiian Planters' Record* 37, no. 3 (1933): 124–162.

18. Ibid.

19. Ibid.

20. Ibid.

21. Ibid.

22. Prinsen Geerligs, *The World's Sugar Cane Industry.*

23. U.S. Department of Commerce, *The Cane Sugar Industry.*

24. Ibid.

25. R. Q. Smith, "History of fertilizer usage in Hawai'i," *Hawaiian Planters' Record* 55 (1955): 55–63; R. P. Humbert, *The Growing of Sugar Cane* (New York: Elsevier Publishing Company, 1968).

26. F. W. Broadbent, "The use of waste molasses as a soil amendment to cane lands," in *Gilmore's Hawaii Sugar Manual 1931* (New Orleans: A. B. Gilmore, 1931), 79–81.

27. H. A. Lee, "A comparison of the root weights and distribution of H 109 and D 1135 cane varieties," *Hawaiian Planters' Record* 33 (1926): 520–523; H. A. Lee, "A comparison of the root distributions of Lahaina and H 109 cane varieties," *Hawaiian Planters' Record* 33 (1926): 523–526; H. A. Lee, Letter to the Director, Experiment Station, HSPA, July 19, 1926, Hawaiian Sugar Planters' Association.

28. U.S. Department of Commerce, *The Cane Sugar Industry.*

29. Ibid.

30. Ibid.

31. Prinsen Geerligs, *The World's Sugar Cane Industry*; U.S. Department of Commerce, *The Cane Sugar Industry*; A&B Sugar Museum.

32. Prinsen Geerligs, *The World's Sugar Cane Industry*.

33. U.S. Department of Commerce, *The Cane Sugar Industry*.

34. Prinsen Geerligs, *The World's Sugar Cane Industry*.

35. U.S. Department of Commerce, *The Cane Sugar Industry*.

36. Ibid.

37. Prinsen Geerligs, *The World's Sugar Cane Industry*.

38. U.S. Department of Commerce, *The Cane Sugar Industry*.

39. Prinsen Geerligs, *The World's Sugar Cane Industry*; U.S. Department of Commerce, *The Cane Sugar Industry*.

40. U.S. Department of Commerce, *The Cane Sugar Industry*.

41. Ibid.

42. Ibid.

43. Ibid.

44. Prinsen Geerligs, *The World's Sugar Cane Industry*.

45. U.S. Department of Commerce, *The Cane Sugar Industry*.

46. Ibid.

47. Ibid.

48. Prinsen Geerligs, *The World's Sugar Cane Industry*.

49. Prinsen Geerligs, *The World's Sugar Cane Industry*; U.S. Department of Commerce, *The Cane Sugar Industry*.

50. W. R. McAllep, "Recent development in factory practice and equipment: Milling," in *Gilmore's Hawaii Sugar Manual 1931* (New Orleans: A. B. Gilmore, 1931), 45–50.

51. Ibid.

52. Ibid.

53. Walter E. Smith, "Recent development in factory practice and equipment: Boiling house," in *Gilmore's Hawaii Sugar Manual 1931* (New Orleans: A. B. Gilmore, 1931), 50–53.

54. Prinsen Geerligs, *The World's Sugar Cane Industry*.

55. U.S. Department of Commerce, *The Cane Sugar Industry*.

56. Smith, "Recent development in factory practice."

57. W. W. Goodale, "Brief history of Hawaiian unskilled labor," in *Hawaiian Almanac and Manual for 1914* (Honolulu: Thos. G. Thrum, 1913), 170–191.

58. Ronald Takaki, *Pau Hana: Plantation Life and Labor in Hawai'i, 1835–1920* (Honolulu: University of Hawai'i Press, 1983).

59. "Mabuhay!" *Ampersand,* Winter 1980–1981; J. W. Vandercook, *King Cane: The Story of Sugar in Hawai'i* (New York: Harper & Brothers Publishers, 1939); Takaki, *Pau Hana.*

60. Thomas G. Thrum, *Hawaiian Almanac and Annual for 1914* (Honolulu: Thos. G. Thrum, 1913).

61. Thomas G. Thrum, *Hawaiian Almanac and Annual for 1922* (Honolulu: Thos. G. Thrum, 1921).

62. A&B Sugar Museum.

63. U.S. Department of Commerce, *The Cane Sugar Industry*.

64. Ibid.

65. Ibid.

66. Ibid.

67. Ibid.

68. Ibid.

69. A&B Sugar Museum.

70. Thrum, *Hawaiian Almanac for 1922.*

71. U.S. Department of Commerce, *The Cane Sugar Industry.*

72. Takaki, *Pau Hana.*

73. U.S. Department of Commerce, *The Cane Sugar Industry;* "American-Hawaiian Steamship Company," http://en.wikipedia.org/wiki/American-Hawaiian_Steamship_Company (accessed April 19, 2012).

74. A. R. Grammer, "A history of the Experiment Station of the Hawaiian Sugar Planters' Association, 1895–1945," *Hawaiian Planters' Record* 51 (1947): 177–228.

75. O. H. Swezey, "Introduction of beneficial insects in Hawai'i," in *Hawaiian Almanac and Manual for 1915* (Honolulu: Thos. G. Thrum, 1914), 128–133.

76. Grammer, "A history."

77. Ibid.

78. U.S. Department of Agriculture, "Establishment and progress of experiment stations," in *Yearbook of Agriculture, 1905* (Washington, DC: U.S. Government Printing Office, 1906).

79. E. V. Wilcox, "Co-operation among farmers," in *Hawaiian Almanac and Manual for 1914* (Honolulu: Thos. G. Thrum, 1913), 154–158.

80. M. M. O'Shaughnessy, "Irrigation," *The Planter's Monthly* 21 (1902): 615–622.

81. Hawai'i Department of Agriculture, *History of Agriculture in Hawai'i.* http://hawaii.gov/hdoa/ag-resources/history.

82. The University of Hawai'i, *History of the University of Hawai'i System.* http://www.hawaii.edu/about/history.html.

83. Hawai'i Department of Agriculture, *History of Agriculture in Hawai'i.*

84. Prinsen Geerligs, *The World's Sugar Cane Industry.*

85. F. G. Krauss, "The Hawaiian homestead of the future," in *Hawaiian Almanac and Manual for 1914* (Honolulu: Thos. G. Thrum, 1913), 158–170.

86. Ibid.

87. Hawai'i Department of Agriculture, *History of Agriculture in Hawai'i.* http://hdoa.hawaii.gov/wp-content/uploads/2013/01/HISTORY-OF-AGRICULTURE-IN-HAWAII.pdf.

88. U.S. Department of Commerce, *The Cane Sugar Industry.*

89. U.S. Department of Agriculture, "Crops other than grains, fruits, and vegetables," *Yearbook of Agriculture, 1924* (Washington, DC: U.S. Government Printing Office, 1925).

90. Alexander & Baldwin, *Ninety Years a Corporation: 1900–1990* (Honolulu: Alexander & Baldwin, Inc., 1990).

91. Alexander & Baldwin. *Ninety Years;* Larry Ikeda, "HC&S Centennial," *Ampersand,* Spring 1982.

92. Kīhei Plantation Company, Manager's Report, December 31, 1903, Hawaiian Sugar Planters' Association; Ikeda, "Centennial."

93. *Maui News,* August 24, 1901; *Maui News,* January 24, 1903; A&B Sugar Museum.

94. W. H. Dorrance, *Sugar Islands: The 165-Year Story of Sugar in Hawai'i* (Honolulu: Mutual Publishing, 2000); Ikeda, "Centennial."

95. Prinsen Geerligs, *The World's Sugar Cane Industry.*

96. Thrum, *Hawaiian Almanac 1914;* U.S. Department of Commerce, *The Cane Sugar Industry.*

97. Dorrance, *Sugar Islands.*
98. Ikeda, "Centennial."
99. U.S. Department of Agriculture, "Field crops other than grain. Table 380." *Yearbook of Agriculture, 1926* (Washington, DC: U.S. Government Printing Office, 1927).
100. Dorrance, *Sugar Islands.*
101. R. Parker, "Hawai'i's largest sugar producer is plantation that started in a business deal," *Coast Banker,* September 1926; J. G. Smith, "Sugar plantations are biggest "small farmers' in Isles," *Honolulu Advertiser,* September 9, 1923.
102. Gilmore, *Hawaii Sugar Manual 1931.*
103. Carol Wilcox, *Sugar Water: Hawai'i's Plantation Ditches* (Honolulu: University of Hawai'i Press, 1996).
104. O'Shaughnessy, "Irrigation."
105. H. T. Stearns and G. A. Macdonald. *Geology and Ground-Water Resources of the Island of Maui, Hawai'i,* Bulletin No. 7 (Honolulu: Hawai'i Division of Hydrography, 1942).
106. Ibid.
107. Ibid.
108. O'Shaughnessy, "Irrigation."
109. Kīhei Plantation Company, Manager's Report.
110. Prinsen Geerligs, *The World's Sugar Cane Industry.*
111. Ibid.
112. Wadsworth, "A historical summary of irrigation."
113. O'Shaughnessy, "Irrigation."
114. Ibid.
115. Ibid.
116. J. H. Foss, "Ditches and ditch lining in connection with sugar cane irrigation," in *Gilmore's Hawaii Sugar Manual 1931* (New Orleans: A. B. Gilmore, 1931), 63–67.
117. Ibid.
118. R. B. Campbell, J. H. Chang, and D. C. Cox. "Evapotranspiration of sugar cane in Hawai'i as measured by in-field lysimeters in relation to climate," *Proceedings of the International Society of Sugar Cane Technologists* 10 (1960): 637–645.
119. Wadsworth, "A historical summary of irrigation."
120. Wilcox, *Sugar Water.*
121. O'Shaughnessy, "Irrigation."
122. *Maui News,* October 5, 1901.
123. Wilcox, *Sugar Water.*
124. *Maui News,* October 5, 1901.
125. Wilcox, *Sugar Water.*
126. Wilcox, *Sugar Water.*
127. J. G. Smith, "Baldwin's dream of banner plantation fulfilled by sons," *Honolulu Advertiser,* September 8, 1923.
128. Smith, "Baldwin's dream."
129. Ikeda, "Centennial."
130. H. R. Shaw, Report to the Director, Experiment Station, HSPA, June 7, 1928.
131. *Honolulu Advertiser,* March 11, 1927.
132. Shaw, Report to the Director; F. E. House, Report to the Director, Experiment Station, Hawaiian Sugar Planters' Association, December 31, 1927.
133. Shaw, Report to the Director.

134. Ibid.

135. O'Shaughnessy, "Irrigation"; *Maui News*. March 30, 1901.

136. Wilcox, *Sugar Water*.

137. Ibid.

138. Ibid.

139. Ibid.

140. Ibid.

141. Ibid.

142. Ibid.

Chapter 4: Depression, War, Federal Legislation, Science, and Technology—1930 to 1969

1. A. B. Gilmore, *The Hawaii Sugar Manual 1951* (New Orleans: A. B. Gilmore, 1951); Aldrich C. Bloomquist, ed., *The Gilmore Hawaii Sugar Manual* (Moorhead, MN: Bloomquist Publications, 1969).

2. Jose Alvarez and Leo C. Polopolus, "The history of U.S. sugar protection," University of Florida AFAS Extension Publication SC019. Available at http://edis.ifas.ufl.edu/sc019 (accessed October 13, 2012).

3. L. D. Baver, "A decade of research progress: 1950–1959," *Hawaiian Planters' Record* 57 (1963): 1–118; Gilmore, *Sugar Manual 1951;* Bloomquist, *The Gilmore Hawaii Sugar Manual.*

4. Baver, "Research progress."

5. Ibid.

6. "Mabuhay!" *Ampersand*, Winter 1980–1981; U.S. Department of Labor, *Labor Conditions in the Territory of Hawai'i, 1929–1930,* Bulletin No. 534 (Washington, DC: U.S. Government Printing Office, 1931).

7. J. W. Vandercook, *King Cane: The Story of Sugar in Hawai'i* (New York: Harper & Brothers Publishers, 1939).

8. E. W. Greene, "Plantation individual piece work and group piece work," in *The Hawaii Sugar Manual 1931* (New Orleans: A. B. Gilmore, 1931).

9. Greene, "Piece work"; U.S. Department of Labor, *Labor conditions.*

10. Greene, "Piece work"; U.S. Department of Labor, *Labor conditions.*

11. U.S. Department of Labor, *Labor conditions.*

12. Ibid.

13. Ibid.

14. G. Daws, *Shoal of Time: A History of the Hawaiian Islands* (Honolulu: University of Hawai'i Press, 1968).

15. Edward D. Beechert, *Working in Hawai'i* (Honolulu: University of Hawai'i Press, 1985).

16. A. R. Grammer, "A history of the Experiment Station of the Hawaiian Sugar Planters' Association: 1895–1945," *Hawaiian Planters' Record* 51 (1947): 177–228.

17. Baver, "Research progress."

18. Ibid.

19. Ibid.

20. Ibid.

21. Ibid.

22. Ibid.

23. R. P. Humbert, *The Growing of Sugar Cane* (New York: Elsevier Publishing Company, 1968).

24. Humbert, *The Growing of Sugar Cane*; A. C. Trouse, Jr., "Recent soil compaction studies," *Proceedings of the Hawaiian Sugar Technologists* 14 (1955): 25–27; A. C. Trouse, Jr., "Deep tillage in Hawai'i," *Soil Science* 88 (1959): 150–158; A. C. Trouse, Jr., "Some effects of soil compaction on the development of sugar-cane roots," *Soil Science* 91 (1961): 208–217; A. C. Trouse, Jr., "Effects of soil compression on development of sugar cane roots," *Proceedings of the International Society of Sugar Cane Technologists* 12 (1967): 137–152.

25. Humbert, *The Growing of Sugar Cane*; Trouse, "Deep tillage."

26. Humbert, *The Growing of Sugar Cane*.

27. Baver, "Research progress."

28. H. F. Clements, "Crop logging sugar cane in Hawai'i," *Better Crops with Plant Food* 32 (1948): 11–18, 45–48; H. F. Clements and T. Kubota, "Internal moisture relations of sugar cane—The selection of a moisture index," *Hawaiian Planters' Record* 46 (1942): 17; H. F. Clements and T. Kubota, "The primary index, its meaning and application to crop management with special reference to sugar cane," *Hawaiian Planters' Record* 47 (1943): 257–297; H. F. Clements and S. Moriguche, "Nitrogen and sugar cane. The nitrogen index and certain quantitative aspects," *Hawaiian Planters' Record* 46 (1942): 163–190; H. F. Clements, "Recent developments in crop logging of sugar cane," *Proceedings of the International Society of Sugar Cane Technologists* 10 (1959): 522–529; H. F. Clements, *Sugarcane Crop Logging and Crop Control: Principles and Practices* (Honolulu: University of Hawai'i Press, 1980); Humbert, *The Growing of Sugar Cane*.

29. A. S. Ayres, "Factors influencing the mineral composition of sugar cane," *Hawaiian Sugar Technologists Reports* 15 (1936): 29–41; W. W. G. Moir, "The plant food problem," *Hawaiian Sugar Technologists Reports* 9 (1930): 175–203.

30. R. P. Humbert and J. P. Martin, "Nutritional deficiency symptoms in sugar cane," *Hawaiian Planters' Record* 55 (1955): 95–102; T. A. Jones and R. O. Humbert, "A spectrographic study of the variations of the nutrient content of sugar cane," *Hawaiian Planters' Record* 55 (1960): 313–317.

31. Jones and Humbert, "A spectrographic study."

32. Baver, "Research progress"; Humbert, *The Growing of Sugar Cane*; Clements, *Sugarcane Crop Logging*.

33. Humbert and Martin, "Nutritional deficiency symptoms"; Baver, "Research progress."

34. Humbert, *The Growing of Sugar Cane*.

35. R. P. Humbert and A. S. Ayres, "The use of aqua ammonia in the Hawaiian sugar industry," *Proceedings of the International Society of Sugar Cane Technologists*, 9th Congress (1956), pp. 524–538; R. P. Humbert and A. S. Ayres, "The use of aqua ammonia in the Hawaiian sugar industry: ii. Injection studies," *Proceedings of the Soil Science Society of America* 21 (1957): 312–316; R. P. Humbert and J. A. Silva, "Aqua ammonia and sugar yields," *Hawaiian Sugar Planters' Association Special Release* 156 (1957).

36. Baver, "Research progress."

37. R. J. Borden, "Nitrogen effects upon the yield and composition of sugar cane," *Hawaiian Planters' Record* 52 (1948): 1–51.

38. Humbert, *The Growing of Sugar Cane*.

39. Baver, "Research progress"; Humbert, *The Growing of Sugar Cane*.

40. Baver, "Research progress"; Humbert, *The Growing of Sugar Cane*.

41. Baver, "Research progress"; Clements, *Sugarcane Crop Logging*.

42. Baver, "Research progress"; Humbert, *The Growing of Sugar Cane*.

43. Humbert, *The Growing of Sugar Cane.*

44. Baver, "Research progress"; Humbert, *The Growing of Sugar Cane.*

45. W. N. Reynolds and W. Gibson. "Water utilization. III. Sugarcane irrigation in Hawai'i," *Proceedings of the International Society of Sugar Cane Technologists* 13 (1968): 55–67.

46. L. D. Baver, "The meteorological approach to irrigation control," *Hawaiian Planters' Record* 54 (1954): 291–298; U. K. Das, "A suggested scheme of irrigation control using the day-degree system," *Hawaiian Planters' Record* 40 (1936): 109–111; L. G. Nickell, "Water utilization. I. Basic plant-water studies," *Proceedings of the International Society of Sugar Cane Technologists* 13 (1968): 38–48.

47. A. H. Cornelison and R. P. Humbert, "Irrigation interval control in the Hawaiian sugar industry," *Hawaiian Planters' Record* 55 (1960): 331–343.

48. Ibid.

49. Baver, "Meteorological approach"; Cornelison and Humbert, "Irrigation interval control"; Hawaiian Sugar Planters' Association, "Irrigation Research," *Hawaiian Planters' Record* 57 (1963): 60–67; F. E. Robinson, "Using standard weather bureau pans for irrigation interval control," *Hawaiian Sugar Technologists Reports* (1962): 105–110.

50. R. B. Campbell, J. H. Chang, and D. C. Cox, "Evapotranspiration of sugar cane in Hawai'i as measured by in-field lysimeters in relation to climate," *Proceedings of the International Society of Sugar Cane Technologists* 10 (1960): 637–645.

51. F. E. Robinson, "Use of neutron meter to establish soil moisture storage and soil moisture withdrawal by sugarcane roots," *Hawaiian Sugar Technologists Reports* (1962): 206–208.

52. J. H. Chang, "Microclimate of sugar cane," *Hawaiian Planters' Record* 61 (1961): 195–225; Robinson, "Using weather bureau pans."

53. Campbell, Chang, and Cox, "Evapotranspiration"; J. H. Chang, R. B. Campbell, H. W. Brodie, and L. D. Baver, "Evapotranspiration research at the HSPA Experiment Station," *Proceedings of the International Society of Sugar Cane Technologists* 12 (1965): 10–24; P. C. Ekern, "Use of water by sugarcane in Hawai'i measured by hydraulic lysimeters," *Proceedings of the International Society of Sugar Cane Technologists* 14 (1971): 805–811.

54. J. H. Chang, "The role of climatology in the Hawaiian sugar-cane industry: An example of applied agricultural climatology in the tropics," *Pacific Science* 17 (1963): 379–397.

55. C. A. Jones, "A review of evapotranspiration studies in irrigated sugarcane in Hawai'i," *Hawaiian Planters' Record* 59 (1980): 195–214.

56. Baver, "Research progress."

57. Jones, "Review of evapotranspiration."

58. W. Gibson, "A method of determining drip irrigation water application efficiency and deficit," *Hawaiian Sugar Technologists Reports* (1977): 49–52.

59. Reynolds and Gibson, "Water utilization."

60. Ibid.

61. R. D. Gerner, "Continuous long-line system of irrigation," *Hawaiian Planters' Record* 54 (1954): 241.

62. H. W. Hansen, "The level ditch system of irrigation at Oahu Sugar Company," *Hawaiian Planters' Record* 54 (1954): 227.

63. Reynolds and Gibson, "Water utilization."

64. C. M. Vaziri and W. Gibson, "Subsurface and drip irrigation for Hawaiian sugarcane," *Hawaiian Sugar Technologists Reports* (1972): 18–22.

65. Humbert, *The Growing of Sugar Cane.*

66. Vandercook, *King Cane.*

67. Humbert, *The Growing of Sugar Cane;* Vandercook, *King Cane.*

68. Vandercook, *King Cane.*

69. Humbert, *The Growing of Sugar Cane.*

70. Vandercook, *King Cane.*

71. Ibid.

72. Reynolds and Gibson, "Water utilization."

73. Baver, "Research progress.

74. Baver, "Research progress"; R. Forbes, "The 'V' cutter," *Hawaiian Sugar Technologists Reports* 15 (1956): 120–121; Humbert, *The Growing of Sugar Cane;* R. P. Humbert and J. H. Payne, "Losses from wet weather harvesting," *Hawaiian Planters' Record* 55 (1960): 301–311.

75. Vandercook, *King Cane.*

76. Ibid.

77. Ibid.

78. Ibid.

79. Ibid.

80. Ibid.

81. R. E. Beiter, "Mechanical dischargers for low-grade centrifugals," *The Hawaii Sugar Manual 1939* (New Orleans: Gilmore Publishing Co. 1939); Vandercook, *King Cane.*

82. Baver, "Research progress."

83. Humbert, *The Growing of Sugar Cane;* E. C. Watt, "Dirty cane costs Hawai'i $11.02 per ton of sugar," *Cane Transport News* 3 (1960): 3.

84. H. T. Stearns and G. A. Macdonald, *Geology and Ground-Water Resources of the Island of Maui, Hawai'i,* Bulletin No. 7 (Honolulu: Hawai'i Division of Hydrography, 1942).

85. Ibid.

86. Ibid.

87. Ibid.

88. Ibid.

89. Ibid.

90. Ibid.

91. Dorrance, *Sugar Islands.*

92. Ibid.

93. Gilmore, *Sugar Manual 1939*

94. Gilmore, *Sugar Manual 1931, 1939, 1951;* Bloomquist, *The Gilmore Hawaii Sugar Manual.*

95. Gilmore, *Sugar Manual 1931.*

96. Gilmore, *Sugar Manual 1931, 1939.*

97. Bloomquist, *The Gilmore Hawaii Sugar Manual.*

98. Gilmore, *Sugar Manual 1951.*

99. Bloomquist, *The Gilmore Hawaii Sugar Manual.*

100. Dorrance, *Sugar Islands.*

101. Gilmore, *Sugar Manual 1931, 1939, 1951.*

102. Gilmore, *Sugar Manual 1931.*

103. Gilmore, *Sugar Manual 1939.*

104. W. P. Naquin, Jr., "The herringbone system of irrigation," *Hawaiian Planters' Record* 54 (1954): 333; Vandercook, *King Cane.*

105. Gilmore, *Sugar Manual 1939.*

106. Gilmore, *Sugar Manual 1951.*

107. Ibid.

108. Bloomquist, *The Gilmore Hawaii Sugar Manual.*

109. Ibid.

110. Gilmore, *Sugar Manual 1939.*

111. Gilmore, *Sugar Manual 1931.*

112. Gilmore, *Sugar Manual 1939.*

113. Ibid.

114. Gilmore, *Sugar Manual 1931.*

115. Gilmore, *Sugar Manual 1939.*

116. Gilmore, *Sugar Manual 1931, 1939.*

117. Gilmore, *Sugar Manual 1951*; Bloomquist, *The Gilmore Hawaii Sugar Manual.*

118. Gilmore, *Sugar Manual 1931.*

119. R. E. Hughes, "Hawaiian plantation electric power and pumping systems—for Hawaiian Commercial & Sugar Co., Ltd.," in *The Hawaii Sugar Manual 1931* (New Orleans: A. B. Gilmore, 1931), 67–69.

120. Gilmore, *Sugar Manual 1939.*

121. Ibid.

122. Gilmore, *Sugar Manual 1951.*

123. Ibid.

124. Ibid.

125. Gilmore, *Sugar Manual 1951*; Bloomquist, *The Gilmore Hawaii Sugar Manual.*

126. Bloomquist, *The Gilmore Hawaii Sugar Manual.*

127. Gilmore, *Sugar Manual 1931.*

128. Gilmore, *Sugar Manual 1939.*

129. Gilmore, *Sugar Manual 1951.*

130. Ibid.

131. Ibid.

132. Bloomquist, *The Gilmore Hawaii Sugar Manual.*

133. Gilmore, *Sugar Manual 1931.*

134. Gilmore, *Sugar Manual 1939.*

135. Gilmore, *Sugar Manual 1951.*

136. Ibid.

137. Bloomquist, *The Gilmore Hawaii Sugar Manual.*

138. Gilmore, *Sugar Manual 1931.*

139. Gilmore, *Sugar Manual 1939.*

140. Gilmore, *Sugar Manual 1951.*

141. Gilmore, *Sugar Manual 1931.*

142. Gilmore, *Sugar Manual 1939.*

143. Ibid.

144. Gilmore, *Sugar Manual 1951*; Larry Ikeda, "IIC&S Centennial," *Ampersand* (Spring 1982).

145. Gilmore, *Sugar Manual 1951.*

146. Bloomquist, *The Gilmore Hawaii Sugar Manual.*

147. Ibid.

148. Gilmore, *Sugar Manual 1931.*

149. Gilmore, *Sugar Manual 1939.*

150. Ibid.

151. Gilmore, *Sugar Manual 1951*.

152. Ibid.

153. Gilmore, *Sugar Manual 1951*; Bloomquist, *The Gilmore Hawaii Sugar Manual*.

154. Bloomquist, *The Gilmore Hawaii Sugar Manual*.

155. Hughes, "Electric power and pumping systems."

156. Gilmore, *Sugar Manual 1939*.

157. Gilmore, *Sugar Manual 1951*.

158. Bloomquist, *The Gilmore Hawaii Sugar Manual*.

159. Gilmore, *Sugar Manual 1931*.

160. Gilmore, *Sugar Manual 1939*.

161. Ibid.

162. Ikeda, "HC&S Centennial."

163. Ibid.

164. Ikeda, "HC&S Centennial"; Alexander & Baldwin, *Ninety Years a Corporation: 1900–1990* (Honolulu: Alexander & Baldwin, Inc., 1990).

165. Gilmore, *Sugar Manual 1951*.

166. Ikeda, "HC&S Centennial."

167. Ibid.

168. Bloomquist, *The Gilmore Hawaii Sugar Manual*.

169. Ibid.

Chapter 5: Drip Irrigation and New Disease Resistant Varieties—1970–2014

1. Jose Alvarez and Leo C. Polopolus, *The History of U.S. Sugar Protection*, University of Florida AFAS Extension Publication SC019. Available at http://edis.ifas.ufl.edu/sc019 (accessed October 13, 2012).

2. Ibid.

3. W. H. Dorrance, *Sugar Islands: The 165-Year Story of Sugar in Hawai'i* (Honolulu: Mutual Publishing, 2000).

4. Larry Ikeda, "USDA official cites sugar difficulties," *Maui Today*, December 1986; Alvarez and Polopolus, *History of U.S. Sugar Protection;* Don J Heinz and Robert V. Osgood, "A history of the Experiment Station, Hawaiian Sugar Planters' Association: Agricultural progress through cooperation and science," *Hawaiian Planters' Record* 61, no. 3 (2009): 1–106; U.S. Department of Agriculture (USDA), "World and U.S. sugar and corn sweeten prices. Table 4. U.S. raw sugar price, duty-free paid, New York, monthly, quarterly, and by calendar and fiscal year," Washington, DC: Economic Research Service. Available at http://www.ers.usda.gov/data-products/sugar-and-sweeteners-yearbook-tables.aspx#25442 (accessed February 7, 2013).

5. "Sugar future depends on farm act," *Maui Today*, April 1987; Dorrance, *Sugar Islands*.

6. "Sugar earns 'new' respect," *Ampersand*, Spring 1988.

7. Richard Oliver, "Future of sugar industry is 'promising,'" *Maui Today*, April 1990; Heinz and Osgood, "A history of the Experiment Station."

8. Alvarez and Polopolus, *History of U.S. Sugar Protection;* Remy Jurenas, "Sugar Policy and the 2008 Farm Bill," Congressional Research Service. Available at http://nationalaglawcenter.org/wp-content/uploads/assets/crs/RL34103.pdf (accessed February 7, 2013); USDA, "World and U.S. sugar"; Brazilian Sugarcane Industry Association, *U.S. Sugar Policy*. Available at http://sugarcane.org/global-policies/policies-in-the-united-states/sugar-in-the-united-states (accessed April, 30, 2013).

9. Alvarez and Polopolus, *History of U.S. Sugar Protection;* USDA, "World and U.S. sugar."

10. Dorrance, *Sugar Islands.*

11. Ibid.

12. Heinz and Osgood, "A history of the Experiment Station,"

13. Ibid.

14. Ibid.

15. Ibid.

16. Ibid.

17. Ibid.

18. Ibid.

19. Hawaii Agriculture Research Center, *Annual Report of the Experiment Station,* 2009 and 2010.

20. Heinz and Osgood, "A history of the Experiment Station."

21. C. M. Vaziri, "Test plot installed for study of subsurface irrigation," in *Hawaiian Sugar Planters' Association Experiment Station Annual Report,* 1970, 35–36.

22. C. M. Vaziri and W. Gibson, "Subsurface and drip irrigation for Hawaiian sugarcane," *Hawaiian Sugar Technologists Reports* (1972): 18–22.

23. Ibid.

24. W. Gibson, "Hydraulics, mechanics and economics of subsurface and drip irrigation of Hawaiian sugarcane," *Proceedings of the International Society of Sugar Cane Technologists* 15 (1974): 639–648.

25. Ibid.

26. Ibid; Mark Zeug, "Revolution in the cane fields," *Ampersand,* Winter 1980–1981.

27. H. W. Hilton and A. Teshima, "Mouse and rat damage to drip irrigation tubing," *Hawaiian Sugar Planters' Association Experiment Station Annual Report,* 1974; D. Martin, "Drip irrigation," *Proceedings of the International Society of Sugar Cane Technologists* 15 (1974): 637–638.

28. P. H. Koehler, P. H. Moore, C. A. Jones, A. Dela Cruz, and A. Maretzki, "Response of drip irrigated sugarcane to drought stress," *Agronomy Journal* 74 (1982): 906–911.

29. C. A. Jones, "A review of evapotranspiration studies in irrigated sugarcane in Hawai'i," *Hawaiian Planters' Record* 59 (1980): 195–214; R. J. McKenzie, "Amounts of water versus yields in drip-irrigated sugarcane—detecting water stress by remote sensing," *Hawaiian Sugar Planters' Association Experiment Station Annual Report,* 1980, 11–12; R. J. McKenzie, P. H. Moore, J. B. Carr, and H. S. Ginoza, "Amounts of water versus yields in drip-irrigated sugarcane—effects of water stress on anatomical, morphological, and physiological characteristics," *Hawaiian Sugar Planters' Association Experiment Station Annual Report,* 1980, 13–15; L. Santo and R. P. Bosshart, "Amounts of water versus yields of drip-irrigated sugarcane—preliminary results," *Hawaiian Sugar Planters' Association Experiment Station Annual Report,* 1980, 8–11; L. Santo and R. P. Bosshart, "Amounts of water versus yield relationships of drip-irrigated sugarcane—yield results," *Hawaiian Sugar Planters' Association Experiment Station Annual Report,* 1982, 11–12.

30. K. T. Ingram and H. W. Hilton, "Interactions of N-K fertilization and irrigation on sugarcane development and yield," *Hawaiian Sugar Planters' Association Experiment Station Annual Report,* 1983, 17–18; L. Santo and R. P. Bosshart, "Tillage and water effects on yields of drip-irrigated sugarcane in an Oxisol," *Hawaiian Sugar Technolo-*

gists Reports (1984): FE16–FE18; J. A. Sylvester, "Amounts of water vs. yield on heavy clay soils at Kekaha sugar company," *Hawaiian Sugar Technologists Reports* (1984): A32–A39.

31. Santo and Bosshart, "Tillage and water effects."

32. Alexander & Baldwin, *Ninety Years a Corporation: 1900–1990* (Honolulu: Alexander & Baldwin, Inc., 1990); J. A. Engott, and T. T. Vana, "Effects of agricultural land-use changes and rainfall on ground-water recharge in central and west Maui, Hawaiʻi, 1926–2004," U.S. Geological Survey Scientific Investigations Report 2007–5103, 2007; L. T. Santo, S. Schenck, H. Chen, and R. V. Osgood, *Crop Profile for Sugarcane in Hawaiʻi* (Honolulu: Hawaii Agriculture Research Center, 2000).

33. Santo, Schenck, Chen, and Osgood, "Crop Profile."

34. Ibid.

35. Ibid.

36. Ibid.

37. P. H. Moore, and R. V. Osgood, "Prevention of flowering and increasing sugar yield by application of ethephon (2-chloro ethyl phosphonic acid)," *Journal of Plant Growth Regulation* 8, no. 3 (1989): 205–210; Santo, Schenck, Chen, and Osgood, "Crop Profile."

38. H. W. Hilton, R. V. Osgood, and A. Maretski, "Some aspects of Mon 8000 as a sugarcane ripener to replace Polaris," *Proceedings of the International Society of Sugar Cane Technologists* 29 (1980): 328–338; R. V. Osgood, "The effect of several growth regulators on dry matter partitioning in sugarcane Cv. H59–3775," *Hawaiian Sugar Technologists Reports,* 1981, 100–102; Santo, Schenck, Chen, and Osgood, "Crop Profile."

39. Santo, Schenck, Chen, and Osgood, "Crop Profile."

40. Ibid.

41. Ibid.

42. Ibid.

43. Ibid.

44. Marguerite Rho, "Hawaiian agribusiness," *Ampersand,* Winter 1989–1990.

45. Hawaiʻi Department of Agriculture, *Agricultural Water Use and Development Plan* (Honolulu: Hawaiʻi Department of Agriculture, December 2003, revised December 2004).

46. USDA, State fact sheets, Washington, DC: Economic Research Service. Available at http://www.ers.usda.gov/data-products/state-fact-sheets/state-data.aspx?State FIPS=15&StateName=Hawaii.

47. R. Osgood and N. Dudley, "Comprehensive study of biomass yields for tree and grass crops grown for conversion to energy." Final report of the Hawaiian Sugar Planters' Association to the Hawaiʻi Department of Business and Economic Development and Tourism, 1993. Available on request.

48. Charles M. Kinoshita and Jiachun Zhou, *Siting Evaluation for Biomass-Ethanol Production in Hawaiʻi.* Department of Biosystems Engineering, College of Tropical Agriculture and Human Resources, University of Hawaiʻi at Manoa, October 1999. Available at http://www.nrel.gov/docs/fy00osti/29414.pdf (accessed December 19, 2012).

49. E. Alan Kennett, "Bioenergy opportunities at Gay & Robinson," 2006. Available at http://www.hawaiienergypolicy.hawaii.edu/programs-initiatives/bioenergy -alternative-fuels/_downloads/habw-gay-robinson-kennett.pdf (accessed February 12, 2013).

50. Lee Jakeway, "Sugarcane bioenergy potential." Presentation by HC&S to the Hawai'i Agriculture Bioenergy Workshop, 2006. Available athttp://www.hawaiienergypolicy.hawaii.edu/programs-initiatives/bioenergy-alternative-fuels/_downloads/habw-sugarcane-jakeway.pdf (accessed February 12, 2013).

51. Alexander & Baldwin, *Ninety Years a Corporation;* Zeug, "Revolution."

52. Zeug, "Revolution."

53. Ibid.

54. "Water to the roots," *Ampersand,* Spring 1987; Larry Ikeda, "Farming by remote control," *Ampersand,* Winter 1988–1989.

55. Zeug, "Revolution."

56. Harold R. Somerset, "HC&S survival depends on cost-cutting," *Maui Today,* June 1982.

57. "HC&S produces 214,806 tons of sugar," *Maui Today,* December 1983; "HC&S produces a record 223,414 tons of sugar," *Maui Today,* January 1985; "HC&S produces 219,468 tons of raw sugar," *Maui Today,* February 1986; "HC&S reaches new production record," *Maui Today,* January 1987; "HC&S reports record sugar production," *Maui Today,* February 1988.

58. Hawaiian Commercial & Sugar Co. (HC&S), "Yield decline phenomenon at HC&S." *HC&S Agricultural Research Department,* December 1997.

59. H. P. Agee, Letter to W. S. Nicoll, Assistant Manager, Maui Agricultural Company, November 3, 1927; HC&S, "Yield decline."

60. HC&S, "Yield decline."

61. Ibid.

62. "Water to the roots," *Ampersand;* "HC&S reports record sugar production," *Maui Today;* "Cameron describes Maui labor shortage," *Maui Today,* October 1989; Ikeda, "Farming by remote control"; Gary T. Kubota, "Paia sugar mill closes," Starbulletin.com. Available at http://archives.starbulletin.com/2000/09/27/news/story5.html (accessed March 10, 2013); Marsha Petersen "HC&S wins U. S. Senate productivity award," *Maui Today,* September 1989; Rho, "Hawaiian agribusiness."

63. "Water to the roots," *Ampersand;* "HC&S installs automated weather stations," *Maui Today,* July 1988; "Study fails to find health hazards in cane smoke," *Maui Today,* December 1992; Larry Ikeda, "HC&S wants to be 'a good neighbor,'" *Maui Today,* January 1986; Ikeda, "Farming by remote control"; Rho, "Hawaiian agribusiness."

64. USDA Soil Conservation Service, "Regulation of agricultural burning in the State of Hawai'i." Available at http://www.nrcs.usda.gov/Internet/FSE_DOCU MENTS/nrcs143_008684.pdf (accessed February 10, 2013); Timothy Hurley, "Group forms to end cane burning," *Maui News,* October 27, 1997.

65. Allison Sickle, "Maui's cane burning—how dangerous is it?" *Maui Now,* March 23, 2012. Available at http://mauinow.com/2012/03/23/mauis-cane-burning-how-dangerous-is-it/ (accessed February 13, 2013).

66. Wendy Osher, "HC&S responds to cane burning petition," *Maui Now,* September 26, 2012. Available at http://mauinow.com/2012/09/26/hcs-responds-to-cane-burning-petition/ (accessed February 10, 2013); Rick Volner, Jr., "HC&S responds to 'cane burning dangerous?' article," *Maui Now,* March 26, 2012. Available at http://mauinow.com/2012/09/26/hcs-responds-to-cane-burning-petition/ (accessed February 10, 2013).

67. "Partnering to protect the East Maui watershed," *Ampersand,* Winter 1995–1996.

68. Engott and Vana, "Ground-water recharge."

69. Engott and Vana, "Ground-water recharge"; S. B. Gingrich, "Ground-water availability in the Wailuku area, Maui, Hawai'i," U. S. Geological Survey Scientific Investigations Report 2008–5236, 2008.

70. Alexander & Baldwin, Financial reports, *Annual Reports,* 1999–2012. Available at http://www.alexanderbaldwin.com/investor-relations/financial-reports/ (accessed March 9, 2013).

71. HC&S, "Application by HC&S for water use permit in Nā Wai 'Ehā surface water management are for existing agricultural uses ('Īao-Waikapu fields)." Form SW-UPA-E (Maui: Department of Land and Natural Resources, Commission on Water Resource Management, April 22, 2009).

72. HC&S, "Application for water use permit."

73. James T. Paul, "The August 2000 Hawai'i Supreme Court decision: Comments and excerpts regarding the public trust doctrine." Paul Johnson Park & Niles, Council for Hawai'i's Thousand Friends, 2nd ed., 2001. Available at http://www.hawaii.edu/ohelo/pleadings/WaiholeSummary.pdf (accessed February 11, 2013).

74. Commission on Water Resource Management, "Petitions to amend the interim in-stream flow standard for 27 streams in East Maui." Available at http://www.state.hi.us/dlnr/cwrm/act_Petition27EastMaui.htm (accessed February 11, 2013); Chris Hamilton, "Compromise decided for East Maui streams," *Maui News,* May 26, 2010. Available at http://www.mauinews.com/page/content.detail/id/531872.html (accessed February 11, 2013); Chris Hamilton, "No party in water decision satisfied," *Maui News,* May 27, 2010. Available at http://www.mauinews.com/page/content.detail/id/531912.html (accessed February 11, 2013).

75. HC&S, "Application for water use permit."

76. Clyde W. Nāmu'o, Letter to Laura H. Thielen and Ken C. Kawahara, Commission on Water Resource Management, May 26, 2009.

77. "Wailuku Water Co & HC&S to release water in Waihe'e River and Waiehu Stream" (press release), Hawai'i Department of Land and Natural Resources, August 2, 2010, http://hawaii.gov/dlnr/cwrm/news/2010/nr20100802.pdf (accessed February 11, 2013); Wendy Osher, "High court rules on Maui water rights at Nā Wai 'Ehā," *Maui Now,* August 16, 2012, http://mauinow.com/2012/08/16/high-court-rules-on-maui-water-rights-at-na-wai-eha/ (accessed February 11, 2013); Stu Woo, "A fight for Maui's water," *Wall Street Journal,* March 22, 2010, http://online.wsj.com/article/SB10001424052748704454004575135602403890836.html (accessed February 10, 2013).

78. Hamilton, "Compromise decided"; Hamilton, "No party satisfied."

79. C. L. Cheng, "Measurements of seepage losses and gains, East Maui Irrigation diversion system, Maui, Hawai'i." U.S. Geological Survey Open-File Report 2012–1115 (2012).

80. Alexander & Baldwin, Financial reports.

81. Kubota, "Paia sugar mill closes."

82. Alexander & Baldwin, Financial report, 2012.

83. Wendy Osher, "HC&S puts hundreds of workers on furlough," *Maui Now,* December 12, 2008, http://mauinow.com/2008/12/12/hcs-puts-hundreds-of-workers-on-furlough/ (accessed March 10, 2013).

84. Alexander & Baldwin, Financial reports, 2011, 2012.

85. Hawaiian Commercial & Sugar, "Leading Role for HC&S in Hawai'i Renewable Energy Research, Biofuels—Federal Funding to be Provided." Enhanced Online News, 2010.

86. Ibid.
87. Alexander & Baldwin, Financial report, 2012.
88. Jakeway, "Sugarcane bioenergy potential."
89. USDA, State fact sheets.
90. USDA, State fact sheets.

LITERATURE CITED

Adler, Jacob. *Claus Spreckels: The Sugar King in Hawai'i.* Honolulu: University of Hawai'i Press, 1966.

Agee, H. P. Letter to W. S. Nicoll, Assistant Manager, Maui Agricultural Company, November 3, 1927.

Alexander & Baldwin. *Annual Reports,* 1999–2012. http://www.alexanderbaldwin .com/investor-relations/financial-reports/.

———. *Ninety Years a Corporation: 1900–1990.* Honolulu: Alexander & Baldwin, Inc. 1990.

Alexander, W. D. "A brief history of land titles in the Hawaiian Kingdom," 105–124. In *Hawaiian Almanac and Annual for 1891.* Honolulu: Thos. G. Thrum, 1890.

Alexander, W. P. " 'Duty of water'—A relic of early irrigation terminology." *Hawaiian Planters' Record* 32 (1928): 122–130.

Allen, R. M. "Information for the irrigator." *Hawaiian Planters' Record* 22 (1920): 145–164.

Alvarez, Jose, and Leo C. Polopolus. "The history of U.S. sugar protection." University of Florida AFAS Extension Publication SC019. http://edis.ifas.ufl .edu/sc019.

Ayres, A. S. "Factors influencing the mineral composition of sugarcane." *Hawaiian Sugar Technologists Reports* 15 (1936): 29–41.

Baldwin, D. D. "Lahaina cane." *The Planters' Monthly* 1 (1883): 42–43.

Baver, L. D. "A decade of research progress, 1950–1959." *Hawaiian Planters' Record* 57 (1963): 1–118.

———. "The meteorological approach to irrigation control." *Hawaiian Planters' Record* 54 (1954): 291–298.

Beechert, Edward D. *Working in Hawai'i.* Honolulu: University of Hawai'i Press, 1985.

Beiter, R. E. "Mechanical dischargers for low-grade centrifugals." In *The Hawaii Sugar Manual 1939.* New Orleans: A. B. Gilmore, 1939.

Bird, Isabella L. *The Hawaiian Archipelago.* Project Gutenberg EBook No. 6750, 1875.

Bloomquist, Aldrich C., ed. *The Gilmore Hawaii Sugar Manual.* Moorhead, MN: Bloomquist Publications, 1969.

Borden, R. J. "Nitrogen effects upon the yield and composition of sugarcane." *Hawaiian Planters' Record* 52 (1948): 1–51.

Brazilian Sugarcane Industry Association. *Solutions from Sugarcane*. http://sugar cane.org/global-policies/policies-in-the-united-states/sugar-in-the-united -states.

Broadbent, F. W. "The use of waste molasses as a soil amendment to cane lands," 79–81. In *Gilmore's Hawaii Sugar Annual 1931. New Orleans A. B. Gilm-ore, 1931.

Brodie, H. W. "Consumption of water by sugarcane," 39–40. In *Annual Report, Experiment Station, Hawaiian Sugar Planters' Association*, 1961.

"Cameron describes Maui labor shortage." *Maui Today*, October 1989.

Campbell, Archibald. *A Voyage Round the World from 1806 to 1812, in which Japan, Kamschatka, the Aleutian Islands, and the Sandwich Islands were Visited*. New York: Van Winkle, Wiley & Co, 1817.

Campbell, R. B., J. H. Chang, and D. C. Cox. "Evapotranspiration of sugar cane in Hawai'i as measured by in-field lysimeters in relation to climate." *Proceedings of the International Society of Sugar Cane Technologists* 10 (1960): 637–645.

Chang, J. H. "Microclimate of sugarcane." *Hawaiian Planters' Record* 61 (1961): 195–225.

———. "The role of climatology in the Hawaiian sugar-cane industry: An example of applied agricultural climatology in the tropics." *Pacific Science* 17 (1963): 379–397.

Chang, J. H., R. B. Campbell, H. W. Brodie, and L. D. Baver. "Evapotranspiration research at the HSPA Experiment Station." *Proceedings of the International Society of Sugar Cane Technologists*, 12th Congress (1965): 10–24.

Chang, J. H., R. B. Campbell, and F. E. Robinson. "On the relationship between water and sugar cane yield in Hawai'i." *Agronomy Journal* 55 (1963): 450–453.

Cheng, C. L. *Measurements of Seepage Losses and Gains, East Maui Irrigation Diversion System, Maui, Hawai'i*. U.S. Geological Survey Open-File Report 2012–1115, 2012.

Clements, H. F. "Crop logging sugar cane in Hawai'i." *Better Crops with Plant Food* 32 (1948): 11–18, 45–48.

———. "Recent developments in crop logging of sugarcane." *Proceedings of the International Society of Sugar Cane Technologists*, 10th Congress (1959): 522–529.

———. *Sugarcane Crop Logging and Crop Control: Principles and Practices*. Honolulu: University of Hawai'i Press, 1980.

Clements, H. F., and T. Kubota. "Internal moisture relations of sugarcane—The selection of a moisture index." *Hawaiian Planters' Record* 46 (1942): 17–22.

———. "The primary index, its meaning and application to crop management with special reference to sugarcane." *Hawaiian Planters' Record* 47 (1943): 257–297.

Clements, H. F., and S. Moriguche. "Nitrogen and sugarcane. The nitrogen index and certain quantitative aspects." *Hawaiian Planters' Record* 46 (1942): 163–190.

Commission on Water Resource Management. *Petitions to amend the interim instream flow standard for 27 streams in East Maui*. http://www.state.hi.us /-dlnr/cwrm/act_Petition27EastMaui.htm.

Cook, James. *The Three Voyages of James Cook Round the World Complete in Seven Volumes*, Vol. VI. *Being the Second of the Third Voyage*. London: Longman, Hurst, Rees, Orme, and Brown, 1821. http://play.google.com/books/reader ?id=XmR7dnU3DgUC&printsec=frontcover&output=reader&hl=en).

Cornelison, A. H., and R. P. Humbert. "Irrigation interval control in the Hawaiian sugar industry." *Hawaiian Planters' Record* 55 (1960): 331–343.

Crawford, David L. *Hawai'i's Crop Parade: A Review of Useful Products Derived from the Soil in the Hawaiian Islands, Past and Present*. Honolulu: Advertiser Publishing Company Ltd., 1937.

Daniel, H. E., ed. "A&B, land and sea, one hundred and twenty-five years strong." *Ampersand*, 1995.

Daws, G. *Shoal of Time: A History of the Hawaiian Islands*. Honolulu: University of Hawai'i Press, 1968.

Deenik, J., and A. T. McClellan. *Soils of Hawai'i*. Cooperative Extension Service, College of Tropical Agriculture and Human Resources, University of Hawai'i at Manoa, SCM-20, September 2007.

Department of Foreign Affairs, Kingdom of Hawai'i. "Coffee, the coming staple product." In *The Hawaiian Islands, Their Resources Agricultural, Commercial and Financial*. Honolulu: Hawaiian Gazette Company, 1896.

Dorrance, W. H. *Sugar Islands: The 165-Year Story of Sugar in Hawai'i*. Honolulu: Mutual Publishing, 2000.

Ekern, P. C. "Drip irrigation of sugarcane measured by hydraulic lysimeters, Kunia, Oahu." Water Resources Research Center, Technical Report 109. Honolulu: University of Hawai'i, 1977.

———. "Use of water by sugarcane in Hawai'i measured by hydraulic lysimeters." *Proceedings of the International Society of Sugar Cane Technologists* 14 (1971): 805–811.

Engott, J. A., and Vana, T. T. "Effects of agricultural land-use changes and rainfall on ground-water recharge in central and west Maui, Hawai'i, 1926–2004." U.S. Geological Survey Scientific Investigations Report 2007–5103, 2007.

Ewart, G. Y. "Consumptive use and replenishment standards in irrigation." *Proceedings of the International Society of Sugar Cane Technologists* 12 (1965): 34–51.

Experiment Station, Hawaiian Sugar Planters' Association. "The relation of applied science to sugar production in Hawai'i." *Experiment Station Hawaiian Sugar Planters' Association*, 1915.

Foote, D. E., E. L. Hill, S. Nakamura, F. Stephens. *Soil Survey of the Islands of Kauai, Oahu, Maui, Molokai, and Lanai, State of Hawai'i*. Washington: Soil Conservation Service, August 1972.

Forbes, R. "The 'V' cutter." *Hawaiian Sugar Technologists Reports* 15 (1956): 120–121.

Foss, J. H. "Ditches and ditch lining in connection with sugarcane irrigation." In *Gilmore's Hawaii Sugar Manual 1931*. New Orleans: A. B. Gilmore, 1931.

Gerner, R. D. "Continuous long line system of irrigation." *Hawaiian Planters' Record* 54 (1954): 241.

Gibson, W. "Hydraulics, mechanics and economics of subsurface and drip irrigation of Hawaiian sugarcane." *Proceedings of the International Society of Sugar Cane Technologists* 15 (1974): 639–648.

———. "A method of determining drip irrigation water application efficiency and deficit." *Hawaiian Sugar Technologists Reports* (1977) 49–52.

Gilmore, A. B. *The Gilmore Hawaii Sugar Manual 1951.* New Orleans: A. B. Gilmore Publishing Co., 1951.

———. *The Hawaii Sugar Manual 1931.* New Orleans: A. B. Gilmore Publishing Co., 1931.

———. *The Hawaii Sugar Manual 1939.* New Orleans: A. B. Gilmore Publishing Co., 1939.

Gingrich, S. B. "Ground-water availability in the Wailuku area, Maui, Hawai'i." U. S. Geological Survey Scientific Investigations Report 2008–5236, 2008.

Girvin, James W. *The Master Planter.* Honolulu: Press of Hawaiian Gazette Co., 1910.

Goodale, W. W. "Brief history of Hawaiian unskilled labor." *Hawaiian Almanac and Manual for 1914.* Honolulu: Thos. G. Thrum, 1913.

Grammer, A. R. "A history of the Experiment Station of the Hawaiian Sugar Planters' Association 1895–1945." *Hawaiian Planters' Record 51* (1947): 177–228.

Greene, E. W. "Plantation individual piece work and group piece work." *Gilmore's Hawaii Sugar Manual 1931.* New Orleans: A. B. Gilmore Publishing Co., 1931.

Hamilton, Chris. "Compromise decided for East Maui streams." *Maui News,* May 26, 2010. http://www.mauinews.com/page/content.detail/id/531872.html.

———. "No party in water decision satisfied." *Maui News,* May 27, 2010. http://www.mauinews.com/page/content.detail/id/531912.html.

Hansen, H. W. "The level ditch system of irrigation at Oahu Sugar Company." *Hawaiian Planters' Record 54* (1954): 227.

Hawaii Agriculture Research Center, Annual Report of the Experiment Station. Honolulu: Hawaii Agriculture Research Center, 2009 and 2010.

Hawai'i Department of Agriculture. *Agricultural Water Use and Development Plan.* Honolulu: Hawai'i Department of Agriculture, December 2003.

———. *History of Agriculture in Hawai'i.* http://hawaii.gov/hdoa/ag-resources/-history.

Hawaiian Commercial & Sugar Company (HC&S). "Application by HC&S for water use permit in Nā Wai 'Ehā surface water management for existing agricultural uses ('Īao-Waikapu fields). Form SWUPA-E." Department of Land and Natural Resources. Commission on Water Resource Management. April 22, 2009.

———. "Leading Role for HC&S in Hawai'i Renewable Energy Research, Biofuels—Federal Funding to be Provided." *Enhanced Online News,* 2010. http://eon.businesswire.com/news/eon/20100407007061/en.

———. "Yield decline phenomenon at HC&S." HC&S Agricultural Research Department, December 1997.

Hawaiian Sugar Planters' Association. *Agronomy, Irrigation, Weed Control, Pathology, Varieties and Insects and Rats Report Series.* Honolulu: HSPA, 1961–1999.

———. "Irrigation Research." *Hawaiian Planters' Record 57* (1963): 60–67.

———. *Story of Sugar in Hawai'i.* Honolulu: HSPA, 1929.

"HC&S installs automated weather stations." *Maui Today,* July 1988.

"HC&S produces a record 223,414 tons of sugar." *Maui Today,* January 1985.

"HC&S produces 214,806 tons of sugar." *Maui Today,* December 1983.

"HC&S produces 219,468 tons of raw sugar." *Maui Today,* February 1986.

"HC&S reaches new production record." *Maui Today,* January 1987.

"HC&S reports record sugar production." *Maui Today,* February 1988.

Heinz, Don J, and Robert V. Osgood. "A history of the Experiment Station, Hawaiian Sugar Planters' Association: Agricultural progress through cooperation and science." *Hawaiian Planters' Record* 61, no. 3 (2009): 1–106.

Hilton, H. W., R. V. Osgood, and A. Maretski. "Some aspects of Mon 8000 as a Sugarcane ripener to replace Polaris." *Proceedings of the International Sugar Cane Technologists* 29 (1980): 328–338.

Hilton, H. W., and A. Teshima. "Mouse and rat damage to drip irrigation tubing." *Hawaiian Sugar Planters' Association Experiment Station Annual Report,* 1974.

House, F. E. Report to the Director, Experiment Station, Hawaiian Sugar Planters' Association, December 31, 1927.

"HSPA reports on 'disappointing' year." *Maui Today,* April 1989.

Hughes, R. E. "Hawaiian plantation electric power and pumping systems—for Hawaiian Commercial & Sugar Co., Ltd." In *Gilmore's Hawaii Sugar Manual 1931.* New Orleans: A. B. Gilmore Publishing Co., 1931.

Humbert. R. P. "Field experimentation to correlate soil and plant analysis with yields of cane and sugar." *Hawaiian Planters' Record* 55 (1955): 129–136.

———. *The Growing of Sugar Cane.* New York: Elsevier Publishing Company. New York, 1968.

Humbert, R. P., and A. S. Ayres. "The use of aqua ammonia in the Hawaiian sugar industry." *Proceedings of the International Society of Sugar Cane Technologists Congress* 9 (1956): 524–538.

———. "The use of aqua ammonia in the Hawaiian sugar industry: ii. Injection studies." *Proceedings of the Soil Science Society of America* 21 (1957): 312–316.

Humbert, R. P., and J. P. Martin. "Nutritional deficiency symptoms in sugarcane." *Hawaiian Planters' Record* 55 (1955): 95–102.

Humbert, R. P., and J. H. Payne. "Losses from wet weather harvesting." *Hawaiian Planters' Record* 55 (1960): 301–311.

Humbert, R. P., and J. A. Silva. "Aqua ammonia and sugar yields." *Hawaiian Sugar Planters' Association Special Release* 156, 1957.

Hurley, Timothy. "Group forms to end cane burning." *Maui News,* October 27, 1997.

Ikeda, Larry. "Farming by remote control." *Ampersand,* Winter 1988–1989.

———. "HC&S Centennial." *Ampersand,* Spring 1982.

———. "HC&S wants to be 'a good neighbor.'" *Maui Today,* January 1986.

———. "Return to Puʻunēnē." *Ampersand,* Fall 1983.

———. "USDA official cites sugar difficulties." *Maui Today,* December 1986.

Ingram, K. T., and H. W. Hilton. "Interactions of N-K fertilization and irrigation on sugarcane development and yield." *Hawaiian Sugar Planters' Association Experiment Station Annual Report,* 1983.

Jakeway, Lee. "Sugarcane bioenergy potential." Hawaiian Commercial & Sugar Co., 2006. http://www.hawaiienergypolicy.hawaii.edu/programs-initiatives/-bioenergy-alternative-fuels/_downloads/habw-sugarcane-jakeway.pdf.

Jones, C. A. "A review of evapotranspiration studies in irrigated sugarcane in Hawaii." *Hawaiian Planters' Record* 59 (1980): 195–214.

Jones, T. A., and R. O. Humbert. "A spectrographic study of the variations of the nutrient content of sugarcane." *Hawaiian Planters' Record* 55 (1960): 313–317.

Jurenas, Remy. "Sugar Policy and the 2008 Farm Bill." Congressional Research Service. http://w.ww.nationalaglawcenter.org/assets/crs/RL34103.pdf.

Kennett, E. Alan. "Bioenergy opportunities at Gay & Robinson." Gay & Robinson, Inc., 2006.

Kīhei Plantation Company. Manager's Report. December 31, 1903. Hawaiian Sugar Planters' Association.

Kinoshita, Charles M., and Jiachun Zhou. "Siting Evaluation for Biomass-Ethanol Production in Hawai'i." Department of Biosystems Engineering, College of Tropical Agriculture and Human Resources, University of Hawai'i at Manoa, 1999. http://www.nrel.gov/docs/fy00osti/29414.pdf.

Koehler, P. H., P. H. Moore, C. A. Jones, A. Dela Cruz, and A. Maretzki. "Response of drip irrigated sugarcane to drought stress." *Agronomy Journal* 74 (1982): 906–911.

Krauss, Beatrice H. *Native Plants Used as Medicine in Hawai'i.* Honolulu: Harold L. Lyon Arboretum, University of Hawai'i, 1981. http://library.kcc.hawaii.edu/~soma/krauss/ko.html.

Krauss, F. G. "The Hawaiian homestead of the future," 158–170. In *Hawaiian Almanac and Manual for 1914.* Honolulu: Thos. G. Thrum, 1913.

Kubota, Gary T. "Paia sugar mill closes." Starbulletin.com, September 27, 2000. http://archives.starbulletin.com/2000/09/27/news/story5.html.

Larrison, G. K. "Investigation of Hawai'i's water sources," 81–84. In *Hawaiian Almanac and Manual for 1915.* Honolulu: Thos. G. Thrum, 1914.

Lau, L. S., and J. F. Mink. *Hydrology of the Hawaiian Islands.* Honolulu: University of Hawai'i Press, 2006.

Lee, H. A. "A comparison of the root distributions of Lahaina and H 109 cane varieties." *Hawaiian Planters' Record* 33 (1926): 523–526.

———. "A comparison of the root weights and distribution of H 109 and D 1135 cane varieties." *Hawaiian Planters' Record* 33 (1926): 520–523.

———. Letter to the Director, Experiment Station, Hawaiian Sugar Planters' Association, July 19, 1926.

Lydgate, J. M. "Sandalwood days," 50–56. In *Hawaiian Almanac and Manual for 1916.* Honolulu: Thos. G. Thrum, 1915.

"Mabuhay!" *Ampersand,* Winter 1980–1981.

Martin, D. "Drip irrigation." *Proceedings of the International Society of Sugar Cane Technologists Congress* 15 (1974): 637–638.

McAllep, W. R. "Recent development in factory practice and equipment, milling," 45–50. In *Gilmore's Hawaii Sugar Manual 1931.* New Orleans: A. B. Gilmore Publishing Co., 1931.

McKenzie, R. J. "Amounts of water versus yields in drip-irrigated sugarcane—detecting water stress by remote sensing," 11–12. In *Hawaiian Sugar Planters' Association Experiment Station Annual Report,* 1980.

McKenzie, R. J., P. H. Moore, J. B. Carr, and H. S. Ginoza. "Amounts of water versus yields in drip-irrigated sugarcane—effects of water stress on anatomical morphological, and physiological characteristics," 13–15. In *Hawaiian Sugar Planters' Association Experiment Station Annual Report,* 1980.

Meriwether, Lee. "A plantation in Hawai'i." *Harper's Weekly,* November 10, 1888.

Mink, J. F. *State of the Groundwater Resources of Southern Oahu.* Board of Water Supply, City and County of Honolulu, Hawai'i, 1980.

Moir, W. W. G. "The plant food problem." *Hawaiian Sugar Technologists Reports* 9 (1930): 175–203.

Moore, P. H., and R. V. Osgood. "Prevention of flowering and increasing sugar yield by application of ethephon (2-chloro ethyl phosphonic acid)." *Journal of Plant Growth Regulation* 8, no. 3 (1989): 205–210.

Nagata, Kenneth M. "Early Plant Introductions in Hawai'i." *Hawaiian Journal of History* 19 (1985): 35–61. http://evols.library.manoa.hawaii.edu/bitstream /-handle/10524/127/JL19043.pdf?sequence=1.

Nakuina, E. M. "Ancient Hawaiian water rights," 79–84. In *Thrum's Hawaiian Almanac and Annual for 1894.* Honolulu: Thos. G. Thrum, 1893.

Nāmu'o, Clyde W. Letter to Laura H. Thielen and Ken C. Kawahara. Commission on Water Resource Management, May 26, 2009.

Naquin, W. P., Jr. "The herringbone system of irrigation." *Hawaiian Planters' Record* 54 (1954): 333.

Nickell, L. G. "Water utilization. I. Basic plant-water studies." *Proceedings of the International Society of Sugar Cane Technologists* 13 (1968): 38–48.

Norum, E. M. "The use of high capacity sprinklers for the irrigation of sugarcane by Kohala Sugar Company." *Proceedings of the International Society of Sugar Cane Technologists* 13 (1968): 1443–1463.

Oliver, Richard. "Future of sugar industry is 'promising.'" *Maui Today,* April 1990.

Osgood, R. V. "The effect of several growth regulators on dry matter partitioning in sugarcane Cv. H 59–3775." *Proceedings of the Hawaiian Sugar Technologists* (1981): 100–102.

O'Shaughnessy, M. M. "Irrigation." *The Planters' Monthly* 21 (1902): 615–622.

Osher, Wendy. "HC&S puts hundreds of workers on furlough." *Maui Now,* December 12, 2008. http://mauinow.com/2008/12/12/hcs-puts-hundreds -of-workers-on-furlough/.

———. "HC&S responds to cane burning petition." *Maui Now,* September 26, 2012. http://mauinow.com/2012/09/26/hcs-responds-to-cane-burning -petition/.

———. "High Court Rules on Maui Water Rights at Nā Wai 'Ehā." *Maui Now,* August 16, 2012. http://mauinow.com/2012/08/16/high-court-rules-on -maui-water-rights-at-na-wai-eha/.

"Our forestry problems as seen by Hillebrand in 1856." *Hawaiian Planters' Record* 22 (1920): 143–144.

Parker, R. "Hawai'i's largest sugar producer is plantation that started in a business deal." *Coast Banker,* September 1926.

"Partnering to protect the East Maui watershed." *Ampersand,* Winter 1995–1996.

Paul, James T. "The August 2000 Hawai'i Supreme Court decision: Comments and excerpts regarding the public trust doctrine," 2nd edition, 2001. http://www .hawaii.edu/ohelo/pleadings/WaiholeSummary.pdf.

Perry, Antonio. "Hawaiian water rights," 90–99. In *Hawaiian Almanac and Annual for 1913*. Honolulu: Thos. G. Thrum, 1912.

Petersen, Marsha. "HC&S wins U. S. Senate productivity award." *Maui Today*, September 1989.

Pfeiffer, R. J. "Eighty-five years a corporation. 1900 1985." *Ampersand*, 1985.

Pioneer Mill Company, Lahaina, Maui, 1873–1960. Plantation Archives. Hawaiian Sugar Planters' Association. http://www2.hawaii.edu/~speccoll/p_pio neer.pdf .

Planters' Labor and Supply Company of the Hawaiian Islands. *The Planters' Monthly*, 1883, 1886, 1888.

Prinsen Geerligs, H. C. *The World's Sugar Cane Industry: Past and Present*. Manchester: Norman Rodger, Altrincham, 1912.

Reynolds, W. N., and W. Gibson. "Water utilization. III. Sugarcane irrigation in Hawai'i." *Proceedings of the International Society of Sugar Cane Technologists Congress* 13 (1968): 55–67.

Rho, Marguerite. "Hawaiian agribusiness." *Ampersand*, Winter 1989–1990.

Robinson, F. E. "Use of neutron meter to establish soil moisture storage and soil moisture withdrawal by sugarcane roots," 206–208. In *Hawaiian Sugar Technologists Reports*, 1962.

———. "Using standard weather bureau pans for irrigation interval control," 105–110. In *Hawaiian Sugar Technologists Reports*, 1962.

Robinson, F. E., R. B. Campbell, and J. H. Chang. "Assessing the utility of pan evaporation for controlling irrigation of sugarcane in Hawai'i." *Agronomy Journal* 55 (1963): 444–446.

Rolph, George M. *Something about Sugar*. San Francisco: John J. Newbegin, 1917.

Rotzoll, Kolja, Aly I. El-Kadi, and Stephen B. Gingerich. "Estimating hydraulic properties of volcanic aquifers using constant-rate and variable-rate aquifer tests." *J. American Water Resources Association* 43, no. 2 (2007): 334–345.

Santo, L., and R. P. Bosshart. "Amounts of water versus yield relationships of drip-irrigated sugarcane—yield results," 11–12. In *Hawaiian Sugar Planters' Association Experiment Station Annual Report*, 1982.

———. "Amounts of water versus yields of drip-irrigated sugarcane—preliminary results," 8–11. In *Hawaiian Sugar Planters' Association Experiment Station Annual Report*, 1980.

———. "Tillage and water effects on drip-irrigated sugarcane yields," 51–53. In *Hawaiian Sugar Planters' Association Experiment Station Annual Report*, 1983.

———. "Tillage and water effects on yields of drip-irrigated sugarcane in an Oxisol," FE16-FE18. In *Hawaiian Sugar Technologists Reports*, 1984.

———. "Timing of water deficit," 14–15. In *Hawaiian Sugar Planters' Association Experiment Station Annual Report*, 1984.

Santo, L. T., S. Schenck, H. Chen, and R. V. Osgood. *Crop Profile for Sugarcane in Hawai'i*. Honolulu: Hawaii Agriculture Research Center, 2000. http://www.ipmcenters.org/cropprofiles/docs/hisugarcane.html.

Shaw, H. R. Report to the Director, Experiment Station, HSPA, June 7, 1928.

Sickle, Allison. "Maui's cane burning—how dangerous is it?" *Maui Now*, March 23, 2012. http://mauinow.com/2012/03/23/mauis-cane-burning-how-dan gerous-is-it/.

Smith, E. Butler. "Hawaiian plantation electric power and pumping systems—for Pioneer Mill Co., Ltd.," 70–72. In *Gilmore's Hawaii Sugar Manual 1931*. New Orleans: A. B. Gilmore Publishing Co., 1931.

Smith, J. G. "Baldwin's dream of banner plantation fulfilled by sons." *Honolulu Advertiser,* September 8, 1923.

———. "Sugar plantations are biggest 'small farmers' in Isles." *Honolulu Advertiser,* September 9, 1923.

Smith, R. Q. "History of fertilizer usage in Hawai'i." *Hawaiian Planters' Record* 55 (1955): 55–63.

Smith, Walter E. "Recent development in factory practice and equipment. Boiling house," 50–53. In *Gilmore's Hawaii Sugar Manual 1931*. New Orleans: A. B. Gilmore Publishing Co., 1931.

Soil Survey Division Staff. Soil Survey Manual. U.S. Department of Agriculture Soil Conservation Service, 1993.

Somerset, Harold R. "HC&S survival depends on cost-cutting." *Maui Today,* June 1982.

Souza, W. R., and C. I. Voss. "Analysis of an isotropic coastal aquifer system using variable-density flow and solute transport simulation." *Journal of Hydrology* 92 (1987): 17–41.

Stearns, H. T., and G. A. Macdonald. *Geology and Ground-Water Resources of the Island of Maui, Hawai'i*. Bulletin 7. Hawai'i Division of Hydrography, 1942.

"Study fails to find health hazards in cane smoke." *Maui Today,* December 1992.

"Sugar earns "new" respect." *Ampersand,* Spring 1988.

"Sugar future depends on farm act." *Maui Today,* April 1987.

Swezey, O. H. "Introduction of beneficial insects in Hawai'i," 128–133. In *Hawaiian Almanac and Manual for 1915*. Honolulu: Thos. G. Thrum, 1914.

Sylvester, J. A. "Amounts of water vs. yield on heavy clay soils at Kekaha sugar company," A32–A39. In *Hawaiian Sugar Technologists Reports,* 1984.

Takaki, Ronald. *Pau Hana: Plantation Life and Labor in Hawai'i, 1835–1920*. Honolulu: University of Hawai'i Press, 1983.

Tanimoto, T. "4–5 joint as indicators of moisture tension of the sugarcane plant," 265–274. In *Hawaiian Sugar Technologists Reports,* 1961.

Thrum, Thos. G. *Hawaiian Almanac and Annual for 1875*. Honolulu: Thos. G. Thrum, 1874.

———. *Hawaiian Almanac and Annual for 1877*. Honolulu: Thos. G. Thrum, 1876.

———. *Hawaiian Almanac and Annual for 1880*. Honolulu: Thos. G. Thrum, 1879.

———. *Hawaiian Almanac and Annual for 1891*. Honolulu: Thos. G. Thrum, 1890.

———. *Hawaiian Almanac and Annual for 1914*. Honolulu: Thos. G. Thrum, 1913.

———. *Hawaiian Almanac and Annual for 1922*. Honolulu: Thos. G. Thrum, 1921.

Tobin, M. E., R. T. Sugihara, and A. K. Ota. "Rodent damage to Hawaiian sugarcane." In *Proceedings of the 14th Vertebrate Pest Conference,* ed. L. R. Davis and R. E. Marsh. Davis: University of California, 1990.

Trouse, A. C., Jr. "Deep tillage in Hawai'i." *Soil Science* 88 (1959): 150–158.

————. "Effects of soil compression on development of sugarcane roots." *Proceedings of the International Society of Sugar Cane Technologists* 12 (1967): 137–152.

———— "Recent soil compaction studies." *Proceedings of the Hawaiian Sugar Technologists* 14 (1955): 25–27.

————. "Some effects of soil compaction on the development of sugarcane roots." *Soil Science* 91 (1961): 208–217.

U.S. Department of Agriculture. "Crops other than grains, fruits, and vegetables." In *Agriculture Yearbook, 1924*. Washington, DC: U.S. Government Printing Office, 1925.

————. "Establishment and progress of experiment stations in Alaska, Hawaiʻi, and Porto Rico." In *Agriculture Yearbook, 1905*. Washington, DC: U.S. Government Printing Office, 1906.

————. "Field crops other than grain. Table 380." In *Agriculture Yearbook, 1926*. Washington, DC: U.S. Government Printing Office, 1927.

————. "Regulation of agricultural burning in the State of Hawaiʻi." Soil Conservation Service. http://www.nrcs.usda.gov/Internet/FSE_DOCUMENTS/-nrcs143_008684.pdf.

————. "Soil survey laboratory data and descriptions for some soils of Hawaiʻi." Soil Survey Investigations Report. No. 29, Soil Conservation Service, 1976.

————. State Fact Sheets. Economic Research Service. http://www.ers.usda.gov/-data-products/state-fact-sheets/state-data.aspx?StateFIPS=15&StateName=Hawaii.

————. "World and U.S. Sugar and Corn Sweeten Prices. Table 4. U.S. raw sugar price, duty-free paid, New York, monthly, quarterly, and by calendar and fiscal year." Economic Research Service. http://www.ers.usda.gov/data-products/sugar-and-sweeteners-yearbook-tables.aspx#25442.

U.S. Department of Commerce. *The Cane Sugar Industry*. Miscellaneous Series No. 53. Washington, DC: U.S. Government Printing Office, 1917.

U.S. Department of Labor. *Labor Conditions in the Territory of Hawaiʻi, 1929–1930*. Bureau of Labor Statistics, Bulletin 534. Washington, DC: U.S. Government Printing Office, 1931.

Vandercook, J. W. *King Cane: The Story of Sugar in Hawaiʻi*. New York: Harper & Brothers Publishers, 1939.

Vaziri, C. M. "Test plot installed for study of subsurface irrigation," 35–36. In *Hawaiian Sugar Planters' Association Experiment Station Annual Report*, 1970.

Vaziri, C. M., and W. Gibson. "Subsurface and drip irrigation for Hawaiian sugarcane," 18–22. In *Hawaiian Sugar Technologists Reports*, 1972.

Volner, Rick, Jr. "HC&S responds to 'cane burning dangerous?' article." *Maui Now*, March 26, 2012. http://mauinow.com/2012/09/26/hcs-responds-to-cane-burning-petition/.

Wadsworth, H. A. "A historical summary of irrigation in Hawaiʻi." *Hawaiian Planters' Record* 37, no. 3 (1933): 124–162.

"Wailuku Water Co & HC&S to release water in Waiheʻe River and Waiehu Stream." Press Release. Hawaiʻi Department of Land and Natural Resources. August 2, 2010. http://hawaii.gov/dlnr/cwrm/news/2010/nr20100802.pdf.

"Water to the roots." *Ampersand*, Spring 1987.

Watt, E. C. "Dirty cane costs Hawai'i $11.02 per ton of sugar." *Cane Transport News* 3 (1960): 3.

Watt, J. M. "The control of irrigation water." *Hawaiian Planters' Record* 30 (1926): 195–201.

White, L. D. "Canoe Plants of Ancient Hawaii." http://www.canoeplants.com/.

Wikipedia. "American-Hawaiian Steamship Company." http://en.wikipedia.org/wiki/American-Hawaiian_Steamship_Company.

Wilcox, Carol. *Sugar Water: Hawai'i's Plantation Ditches.* Honolulu: University of Hawai'i Press, 1996.

Wilcox, E. V. "Co-operation among farmers," 154–158. In *Hawaiian Almanac and Manual for 1914.* Honolulu : Thos. G. Thrum. 1913.

Wilfong, G. W. "Twenty years' experience in cane culture." *Planters' Monthly* 1 (1882): 147–152.

Woo, Stu. "A fight for Maui's water." *Wall Street Journal,* March 22, 2010. http://online.wsj.com/article/SB10001424052748704454004575135602403890836.html.

Zeug, Mark. "Revolution in the cane fields." *Ampersand,* Winter 1980–1981.

INDEX

Davies Hāmākua Sugar Company, 181, 183, 185, 186
Department of Commerce, 72, 75, 76, 88, 92, 93, 95
diseases (sugarcane): chlorotic streak, 128; downy mildew, 128; eyespot, 163, 200; Fiji, 128; leaf scald, 128, 201; mosaic, 97; pineapple, 128, 200; ratoon stunting, 128, 200; red rot, 128; smut, 128, 185, 186, 200; varietal resistance, 98, 116, 128, 189, 200, 211
ditches: Central, 111, 219, 220; Dole's "water lead," 105; ʻHaikū, 58, 59, 60, 63, 105, 109, 110, 111, 165, 216, 219, 220; Hāmākua, 59, 105, 107, 109, 110; Hanamāʻulu, 104; Honokōhau, 114, 160; Honokōwai, 114, 115; Honolua (Honokōhau), 114, 115; Kahoma, 114, 116; Kaluanui, 110; Kauaʻula, 114, 115; Kauhikoa, 110, 175, 219; Koʻolau, 109, 110, 219, 220; Lahainaluna, 57; Lowrie, 109–111, 175, 219; Launiupoko, 114, 116; Manuel Luis, 111, 219, 220; Olowalu, 114, 116; Rice, 52, 56, 104; South Waiehu, 64; Ukumehame, 114, 116; Waiheʻe, 64, 113, 159, 165, 216, 217; Waiāhole, 217, 218; Wailoa, 110, 111, 165, 175, 216, 219
Dole, James, 99, 100
Dole, Sanford, 39

Eames, Alfred, 99
East Maui Irrigation (company, EMI), 110, 158, 163, 215–217, 219
East Maui Plantation, 27, 29, 30, 37, 46, 54, 60
East Maui Watershed Partnership, 214, 215
electricity (generation), 117, 174, 176, 201–204, 208, 214, 225, 226
Emmeluth, John, 99
Employment Relations Act, Hawaiʻi, 126
ethanol, xvi, 102, 184, 188, 201–205, 214, 220, 222, 226, 228
evaporation pans, 139, 140
ʻEwa Plantation Company, 65, 78, 147, 182, 186

Experiment Station (Hawaiian Sugar Planters' Association), viii, xv, 68, 127, 132, 185, 187, 188
exports: fruits, vegetables, and grains, 11, 13–15, 35, 67, 68, 99, 100, 227; sugar and molasses, 13–15, 19–21, 32, 35, 69, 86, 100, 134; tariffs, 36, 66

Farm Bills, 182, 184, 220, 224
fertilizer: nitrogen (N), 50, 77, 79, 81, 122, 123, 132–136, 170, 171, 193, 196, 197, 211; phosphorus (P), 50, 79, 131–135, 167, 171, 193, 196, 211; potassium (K), 50, 79, 81, 132–136, 167, 193, 196, 205, 209
filter cake, 70, 84, 155
filter presses, 79, 84, 174
flowering (tasseling), 23, 33, 130, 196
flumes, 46, 83, 143, 158
Food and Agriculture Act of 1977, 181
Ford, S. P., 14
fruits and other crops: coffee, 4, 13, 15, 38, 67, 99, 100, 188, 225; pineapple, 4, 14, 97, 99, 100, 126, 128, 135, 156, 163, 171, 191, 200, 202, 204, 208, 225; rice, 14, 15, 17, 40, 66, 67, 78, 100; taro (kalo), 1–3, 5–9, 12–15, 56, 67, 78, 106, 188, 205, 214, 228; other fruits and vegetables (bananas, coconuts, limes, oranges, potatoes, pumpkins, sweet potatoes, tamarinds), 6, 9, 14, 30, 66, 68, 100, 104, 225

Gay & Robinson, 185, 186, 201, 204
George Washington (whaling ship), 24
Ginaca, Henry, 100
Gower, John, 27
groundwater (aquifers), xv, 69, 71, 113, 155–157, 162, 179, 203, 215–217
Grove Ranch Plantation, 30, 37, 54, 64, 158
guano, 50, 69

ʻHaikū Sugar Company, 20, 30, 32, 37, 54, 58, 60, 62–64, 71, 101–103, 109
Hakalau Plantation Company, 57
Hāliʻimaile Plantation, 27
Hāmākua Plantation (Mill Company, Sugar Company), 47, 48, 54, 57, 58, 67, 181, 183, 186, 203

About the Authors

C. Allan Jones is Senior Research Scientist for Texas A&M AgriLife Research. He received a Ph.D. in Botany from Washington State University and has worked as an agricultural research scientist or administrator in Brazil, Colombia, Hawai'i, and Texas. His interest in the Hawai'i sugar industry began in the late 1970s, when he was an agronomist with the Hawaiian Sugar Planters' Association studying sugarcane nutrition and irrigation. He has also served as a plant physiologist with the Agricultural Research Service, as Resident Director of Blackland Research Center, as Associate Director of Texas A&M AgriLife Research, and as Director of the Texas Water Resources Institute. He is author or coauthor of over 90 research publications and five books on the growth and development of tropical grasses, maize, and sugarcane, as well as on mineral nutrition of field crops and agricultural history. He is also a Fellow of the American Society of Agronomy and served as President of the Association for International Agriculture and Rural Development in 2000.

Robert V. Osgood retired in 2003 as Vice President and Assistant Director of Research at the Hawaii Agriculture Research Center (HARC), formerly the Hawaiian Sugar Planters' Association (HSPA). He received M.S. and Ph.D. degrees from the College of Tropical Agriculture at the University of Hawai'i and worked for the Hawai'i sugar industry as an agronomist from 1969 until his retirement in 2003. Beginning in 1982, as a result of the downsizing of the Hawai'i sugar industry, his primary attention was diverted to diversification of Hawai'i's agriculture on the lands coming out of sugarcane. He served as Vice Chair and Chair of the Plant Growth Regulator Society of America (1989–1991). Since his retirement, Robert has served as a consultant for the U.S. Agency for International Development, the Coffee Quality Institute, and for private companies where his work—primarily on sugarcane and coffee production—has taken him to Asia, Africa, Central and South America, and the Caribbean. He has served on several Hawai'i state and nonprofit company boards, including Hawaii Crop Improvement Association, Agribusiness Development Corporation, Hawai'i Agricultural Leaders Foundation, Hawai'i Forest Industry Association, Hawai'i Forest Institute, and O'ahu Resource, Conservation and Development Corporation.